New Mobilities and Social Changes in Russia's Arctic Regions

The world is undergoing the largest wave of urban growth in the history of mankind, and the circumpolar regions are not exempt from this global process. This evolution is linked to the development of industrial activities, as well as to the rise of services, public administration, and tourism. While this trend affects all of the circumpolar countries, it impacts Russia the most. Russia has built a unique urban fabric in the Arctic, with half a dozen cities of between 100,000 and 300,000 inhabitants. Because of their combination of historic, demographic, and economic features, Russian Arctic cities offer a distinctive field to observe and understand issues related to circumpolar urbanization.

This book provides the first in-depth, multidisciplinary study of re-urbanization in Russia's Arctic regions, with a specific focus on new mobility patterns, and the resulting birth of new urban Arctic identities in which newcomers and labor migrants form a rising part. It is an invaluable reference for all those interested in current trends in circumpolar regions, showing how the Arctic is becoming more diverse culturally, but also more integrated into globalized trends in terms of economic development, urban sustainability, and migration.

Marlene Laruelle is Associate Director of the Institute for European, Russian, and Eurasian Studies (IERES) and Research Professor of International Affairs at the Elliott School of International Affairs, The George Washington University, USA.

Routledge Research in Polar Regions

Series Edited by Timothy Heleniak

The Routledge series in *Polar Regions* seeks to include research and policy debates about trends and events taking place in two important world regions, the Arctic and Antarctic. Previously neglected periphery regions, with climate change, resource development, and shifting geopolitics, these regions are becoming increasingly crucial to happenings outside these regions. At the same time, the economies, societies, and natural environments of the Arctic are undergoing rapid change. This series seeks to draw upon fieldwork, satellite observations, archival studies, and other research methods which inform about crucial developments in the Polar regions. It is interdisciplinary, drawing on work from the social sciences and humanities, bringing together cutting-edge research in the Polar regions with the policy implications.

New Mobilities and Social Changes in Russia's Arctic Regions
Edited by Marlene Laruelle

Environment, Resources and Politics in Greenland
Mark Nuttall

New Mobilities and Social Changes in Russia's Arctic Regions

Edited by Marlene Laruelle

Routledge
Taylor & Francis Group

LONDON AND NEW YORK

First published 2017
by Routledge
2 Park Square, Milton Park, Abingdon, Oxon OX14 4RN

and by Routledge
52 Vanderbilt Avenue, New York, NY 10017

First issued in paperback 2020

Routledge is an imprint of the Taylor & Francis Group, an informa business

British Library Cataloguing in Publication Data
A catalogue record for this book is available from the British Library

Library of Congress Cataloging in Publication Data
Names: Laruelle, Marláene, editor.
Title: New mobilities and social changes in Russia's Arctic regions / [edited by] Marlene Laruelle.
Description: Abingdon, Oxon ; New York, NY : Routledge, 2017. |
Series: Routledge research in Polar Regions | Includes bibliographical references.
Identifiers: LCCN 2016011422| ISBN 9781138191471 (hardback) | ISBN 9781315640471 (ebook)
Subjects: LCSH: Russia, Northern—Social conditions. | Russia, Norther—Economic conditions. | Russia, Northern—Emigration and immigration—Economic aspects. | Russia, Northern—Emigration and immigration—Social aspects. | Arctic Coast (Russia)—Economic conditions. | Arctic Coast (Russia)—Social conditions.
Classification: LCC HN530.2.N66 N49 2017 | DDC 306.0947/1—dc23
LC record available at https://lccn.loc.gov/2016011422

ISBN 13: 978-0-367-66816-7 (pbk)
ISBN 13: 978-1-138-19147-1 (hbk)

Typeset in Times New Roman
by diacriTech, Chennai

Contents

Figures

Maps

Tables

Contributors

Editor

Marlene Laruelle is Associate Director of the Institute for European, Russian, and Eurasian Studies (IERES) and Research Professor of International Affairs at the Elliott School of International Affairs, The George Washington University. She previously was a Visiting Scholar at the Woodrow Wilson International Center for Scholars (2005–2006). She holds a PhD from the National Institute for Oriental Languages and Cultures in Paris. She has authored numerous books, including *Russian Eurasianism: An Ideology of Empire* (Johns Hopkins University Press, 2008), *In the Name of the Nation: Nationalism and Politics in Contemporary Russia* (Palgrave Macmillan, 2009), and *Russia's Strategies in the Arctic and the Future of the Far North* (M.E. Sharpe, 2013). She recently edited *Eurasianism and the European Far Right: Reshaping the Russia-Europe Relationship* (Lexington, 2015). She is a co-PI on several NSF grants devoted to Arctic Urban Sustainability.

Contributors

Alexandra Burtseva is Associate Professor at Murmansk State Arctic University. Her field of research includes labor migration on the Kola Peninsula, and issues related to migrants' integration in the Murmansk region. She began working on migrants' integration in 2012, within the framework of the Russian-Finnish project *School for All: Developing Inclusive Education*.

Sophie Hohmann is Associate Researcher at the Centre for Russian, Caucasian and East-European Studies (CERCEC, EHESS/CNRS) and at the Fondation des Sciences de l'Homme in Paris. She is also Research Fellow on the European Project Cascade (Security and Democracy in the Neighbourhood: The Case of the Caucasus). She holds a PhD in social sciences and demography from the EHESS (School of Advanced Studies in Social Sciences) in Paris. Her current research focuses on demographic trends and migration issues in and from South Caucasus and Central Asia to Russia's Arctic cities. She recently co-edited *Development in Central Asia and the Caucasus: Migration, Democratisation and Inequality in the Post-Soviet Era* (I.B. Tauris, 2014).

Natalia Krasnoshtanova, Candidate of Geographical Sciences, is a researcher at the V.B. Sochava Institute of Geography of the Siberian Branch of the Russian Academy of Science, Irkutsk. She has published several articles in Russian on social-economic geography and the ecology of the oil and gas extractive activity regions of Siberia.

Vera Kuklina is a Post-Doctoral Student at the V.B. Sochava Institute of Geography of the Siberian Branch of the Russian Academy of Sciences, Irkutsk, and a Visiting Scholar at the Institute for European, Russian, and Eurasian Studies, The George Washington University. Her research interests are in the field of cultural and social geography of remote regions, urban sustainability, ethnic minorities of Siberia, and social impact assessment of industrial projects. Her publications include *Local Communities in Multiethnic Environment of South Siberia: A Cultural-Geographical Perspective* (Novosibirsk: Siberian Branch of RAS, 2006 [In Russian]) and "Construction of Homeland among Buriats in Irkutsk" (2012, *Sibirica* 12[2]). Her current research interests are cultural and social sustainability issues, post-Soviet legacies, and housing and infrastructure of peripheral communities of Siberia.

Genevieve Parente is a doctoral candidate in geography at the University of British Columbia, Vancouver. Her dissertation examines the role of the private sector in Arctic governance and economic development. Parente has conducted field research in both the U.S. and Russian Arctic. Prior to beginning her PhD, she worked in Washington, DC, for the U.S. Department of Labor on corporate social responsibility issues, and the U.S. Department of State on international labor and migration. During this time, she also earned a certificate in international migration studies from the Institute for the Study of International Migration (ISIM) at Georgetown University. Parente holds an MA in geography from The George Washington University and a BA in history from the University of Virginia. Her MA thesis examines the municipal strategies of two Siberian industrial cities to achieve sustainability in an emerging market economy.

Nikolay Petrov is a professor at the Higher School of Economics in Moscow. For many years he was a scholar-in-residence at the Carnegie Moscow Center, where he directed the Society and Regions project. He also heads the Center for Political-Geographic Research. Petrov is a columnist for the *Moscow Times*, a member of the Program on New Approaches to Research and Security in Eurasia (PONARS Eurasia), and a member of the editorial boards of *Journal of Power Institutions in Post-Soviet Societies, Russian Politics and Law*, and *Region: Regional Studies of Russia, Eastern Europe, and Central Asia*. From 1990 to 1995, he served as an advisor to the Russian parliament, government, and presidential administration. He has published numerous studies analyzing Russia's political regime, post-Soviet transformation, socioeconomic and political development of Russia's regions, democratization, federalism, and elections, among other topics. His works include the three-volume *1997 Political Almanac of Russia* and its annual supplements. He is the coauthor and editor of *The Dynamics of Russian Politics: Putin's Reform of Federal-Regional Relations* in two volumes (2004, 2005), *Irregular Triangle: Interrelations between Authorities, Business, and Society in Russian Regions* (2010), *Russia in 2020: Scenarios for the Future* (2011), and *Russia 2025: Scenarios for the Russian Future* (Palgrave Macmillan, 2013).

Olga Povoroznyuk is a Researcher at the Department for Social and Cultural Anthropology, University of Vienna, and a Senior Researcher at the Department of Northern and Siberian Studies, Institute of Ethnology and Anthropology, Russian Academy of Sciences in Moscow. She has conducted long-term anthropological research on land use practices, culturescapes, gender relations, ethnicity, and identity politics in indigenous and mixed communities in East Siberia, the Russian Far East, and the Nenets district. This research has yielded a number of papers and a book on Soviet history and post-Soviet sociocultural transformations among Evenki communities in the Northern Transbaikal region. Dr. Povoroznyuk currently works on a project devoted to socialities, mobilities, and identity construction processes involving multicultural population of "ethnic" villages and urbanized settlements located along the Baikal-Amur Mainline in the context of the railroad's construction history and current infrastructural and community development.

Gertrude Saxinger is Assistant Professor at the Department of Social and Cultural Anthropology, University of Vienna, as well as at the Austrian Polar Research Institute (APRI). She is also an Adjunct researcher at Yukon College in Whitehorse, Canada. Her field of research is extractive industries in the Arctic and beyond, labor studies in the context of mobility and multi-locality (FIFO and LDC), as well as industry–community relations. She published *Unterwegs – Mobiles Leben in der Erdgas- und Erdölindustrie in Russlands Arktis (Lives on the Move – mobile lives in the Arctic Russian petroleum industry)* (Böhlau Publishers, 2016).

Denis Sokolov began researching social and economic issues of the Northern Caucasus in 2008 while heading the North Caucasus projects at the Moscow School of Political Studies. He later joined the Gaidar Institute for Economic Policy as a research fellow. From 2009 until 2014, he headed the Center for Social and Economic Research of Regions (RAMCOM), which specializes in the Northern Caucasus. Since 2011 he had focused on migration flows from Dagestan to the Western Siberia and Arctic regions. Sokolov is currently a senior research fellow at the Russian Presidential Academy of National Economy and Public Administration (RANEPA) and the director of research at the Center for Social and Economic Research of Regions in Moscow.

Aimar Ventsel is a Guest Lecturer at the Department of Ethnology at the University of Tartu, Estonia, and a visiting scholar at the Department of Sociology of the University of Warwick, UK. He was a founding member of the Siberia Research Group at the Max Planck Institute for Social Anthropology in Halle. He received his PhD in 2005 from Martin Luther University Halle-Wittenberg. His academic interests are identity and property relations in the East Siberian music business, nomadism, and subcultures in Siberia and Germany. His publications include *Reindeer, Rodina, and Reciprocity: Kinship and Property Relations in a Siberian Village* (Lit, 2005); "Punx and Skins United: One Law for Us, One Law for Them" (2008, *Journal of Legal Studies* 57: 45–100); "Consumption and Popular Culture among Youth in Siberia" (with J.O. Habeck, *Zeitschrift für Ethnologie* 134 [2009]: 1–22); and "Generation P in the Tundra: Youth in Siberia" (*Special Issue of Electronic Journal of Folklore* 41, 2009).

Aleksey Yashunsky is Head of the theoretical cybernetics section at RAS Keldysh Institute for Applied Mathematics in Moscow. His interests in geography include GIS, statistical methods, and big data usage in human geography studies.

Nadezhda Zamyatina is a Lead Researcher at Moscow State University. She works on regional identity and cultural geography and studies endogenous factors of local development. She is an Expert at the Council for Productive Forces Research, under the Ministry of Economic Development and Russian Academy of Sciences, where she takes part in municipal and regional strategic planning. She recently published *Rossiia, kotoruiu my obreli. Issleduia prostranstvo na mikrourovne* (Moscow, 2013), with Alexander Pelyasov.

Introduction

Marlene Laruelle

The world is undergoing the largest wave of urban growth in the history of mankind. Half of the world's population now lives in urban areas. The United Nations predicted that by 2050, 64.1 percent and 85.9 percent of the developing and developed world, respectively, will be urbanized (United Nations 2014). The circumpolar regions are not exempt from this global process. Since the 1960s, most of the population growth in the Arctic has occurred in urban centers (Rasmussen 2011, 22). Today, with the exception of the Faroe Islands, all of the Arctic regions have three-quarters or more of their populations residing in urban areas (Larsen and Fondahl 2015, 94). This evolution is linked to the growth in industrial activities (mostly large-scale fishing, forestry, and mineral extraction), but also to the development of social services, public administration, and tourism. For many Arctic residents, only urban centers offer decent living conditions, even if it means, for indigenous people, cutting ties with some of their traditional ways of life.

While this trend affects all of the circumpolar countries, it impacts Russia the most. Indeed, Russia's efforts to build urban settlements in the Far North date to the seventeenth century, far earlier than neighboring countries. Beginning in the 1930s, the Soviet regime promoted Arctic development strategies that were based on permanent (sometimes forced) urban and industrialized settlements. Other Arctic countries did not pursue urbanization until the 1960s. Russia has long been the most urban circumpolar region in the world, growing from a few tens of thousands of inhabitants in 1926 to 1.3 million in 1959 and 2.6 million in 1989 (Josephson 2014, 241). Today Russia hosts about two-thirds of the Arctic population: about 2.5 million in 2015, to which should be added about ten million inhabitants living in sub-Arctic conditions in Siberia (see Map I.1).

Russia has built a unique urban fabric in the Arctic, with half a dozen cities with a population between 100,000 and 300,000 people. Most of Russia's 16 Arctic regions have between 66 and 92 percent of the population classified as urban, as can be seen in Map I.1 (Perepis' 2010). Moreover, Russia's economy depends heavily on the Arctic: around 20 percent of the country's GDP and exports are generated north of the Arctic Circle (Laruelle 2014). Because of this combination of historic, demographic, and economic features, Russian Arctic cities offer a distinctive field to observe and understand issues related to circumpolar urbanization.

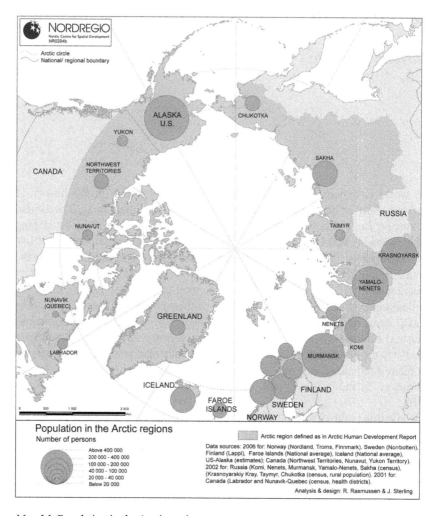

Map I.1 Population in the Arctic regions.

With the collapse of the Soviet Union in 1991, all of Russia's regions underwent a drastic transition from central planning to a market economy. Social reforms greatly expanded the freedom of movement and the right to emigrate, substantially reducing the population ready to live in climatically harsh regions. Between 1987 and 2000 economic output fell by four-fifths in Sakha-Yakutiya and Chukotka. Some mining centers and industrial settlements were totally abandoned, while several military bases were closed. The downsizing of the Northern bonus packages offered by the Soviet state to Arctic inhabitants accelerated the exodus of population. The absence of work prospects; lack of a future for their children; exorbitant prices for basic goods; chronic shortages of heating, gas, and electricity; and the declining transportation links with the rest of the country have pushed millions of Russians to migrate to the European regions of the country (Heleniak 2009a). The majority migrated outside of any state-organized framework. Between the censuses of 1989 and 2002, the regions of Magadan and Chukotka lost more than half of their populations, Taimyr 30 percent, Yamalo-Nenets 25 percent, and even the Murmansk region lost more than 20 percent of its population. Sakha-Yakutiya escaped relatively lightly, with a depopulation of only 12 percent (Heleniak 1999). As noted by Timothy Heleniak, between 1993 and 2009 the High North "had a population decline of 15.3 percent, consisting of 17.1 percent decline from net out-migration, compensated for by a 1.8 percent increase from the region having more births than deaths as a result of having a younger age structure than the country" (Heleniak 2010, 17–18).

Today's Russian Federation is a fragmented territory in terms of population, access to wealth, human development indicators, and economic strategies. Within the space of two decades it has become a de facto archipelago, in which social inequalities are above all regional inequalities. While some modern and wealthy "islands" are developing across its immense landmass, other areas are being emptied of their populations, economically impoverished, and increasingly disconnected from the rest of the country. "Metropolitan Russia" – Moscow, St. Petersburg, Yekaterinburg, Nizhny-Novgorod, Kazan – is distinguished by its high level of revenue, inhabitants with higher education, and generous access to services. The university and science towns of Siberia can also be added to this list, such as Novosibirsk, Omsk, Tomsk, and Krasnoyarsk, which have lower revenues but a high degree of access to the outside world. A second, "rent archipelago," Russia – Tyumen, Surgut, and Khanty-Mansiysk – has the highest revenues per capita in the country and offers its inhabitants generous social policies and broad access to technologies. A relatively similar situation concerns also, to a lesser extent, several other Arctic cities based on extraction, from Novyi Urengoi to Norilsk. The Black Earth – situated between Kursk, Tambov, Volgograd, and Krasnodar – is the only region to record both economic and demographic growth. The rest of the country can be defined as "second-class Russia," characterized by abandoned industrial towns in full crisis, high unemployment rates, the pauperization of the former Soviet middle classes, agricultural wastelands, very poor access to transport, and acute demographic crisis (Dienes 2002; Zurabevich ongoing research).

However, since the 2000s, the focus on developing Arctic resources by both the Russian state and private actors (domestic as well as foreign) gave new impetus for many people to migrate to the region, prompting a renewed vibrancy for some Northern cities. With the exception of Moscow and the Moscow region, the main Arctic and sub-Arctic regions, well endowed with resources, display a high gross regional product (GRP) per capita, with Tyumen leading the group (see Figure I.1).

Economics is, therefore, at the root of Russia's Arctic demographic trends: as shown in Figure I.2, Arctic regions face an outflow of young people going to study in other cities in the country, and another outflow of people of retirement age moving toward more southern regions. But they also receive an inflow of working age people – mostly between 20 and 35 – who want to build professional careers in the Arctic for themselves and their family.

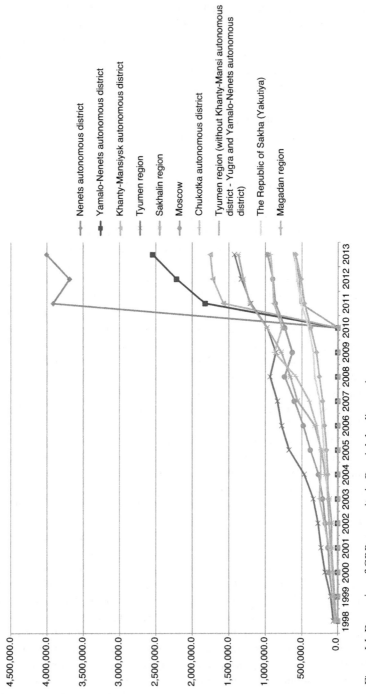

Figure 1.1 Dynamics of GRP per capita in Russia's leading regions.

Source: Complied from Rosstat for every year.

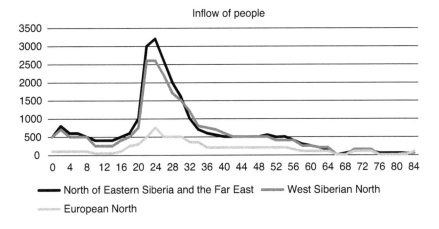

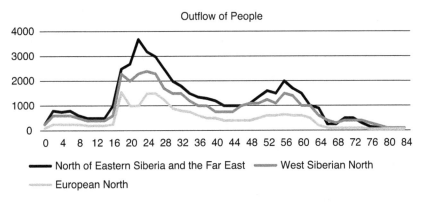

Figure I.2 Age distribution of arrivals and departures in regions of the Far North, 2010.

Source: Igor Efremov, "Vozrastnye osobennosti migratsionnykh protsessov na Krainem Severe Rossii," *Demoscope*, 2014, http://demoscope.ru/weekly/2014/0581/analit06.php.

The Arctic regions thus display a high mobility or, more precisely, several contradictory mobilities: out and in for emigration/immigration flows; south–north in terms of economic flows between regional centers, traditionally located in southern Siberia, and Northern cities; north–north if we follow the professional trajectories of those Northerners building their career in several Arctic cities over the years. New trends of mobility also emerge: indigenous populations leaving rural, impoverished regions to reach a bigger town or the regional capital, long-distance commuters following the new oil and gas exploration trends, and labor migrants from other post-Soviet countries in search of job opportunities.

Many of Russia's Arctic cities are now part of a globalized job market, attracting newcomers either from the indigenous populations or from far more distant regions and countries. This new trend has been noted by the 2014 *Arctic Human Development Report*, which states that "new flows of immigrants to the Arctic and related cultural negotiations and contestations – in short the magnitude and complexity of migration and urbanization – pose multifaceted challenges to human development in the Arctic" (Larsen and Fondahl 2015, 22). Yet, research on these new mobility patterns remains scarce.

These new mobilities have a complex impact on indigenous populations. Russia's unique feature in the circumpolar ethnic landscape is that indigenous groups only constitute a very small percentage of the total figure of Arctic inhabitants. Indigenous peoples represent 80 percent of Greenland's population, 50 percent of Canada's, 20 percent of Alaska's, and 15 percent of Arctic Norway's, but they make up less than 5 percent of Arctic Russia (see Map I.2). However, they have a more solid demography than ethnic Russians and therefore have seen their share of the Arctic population increase over these last two decades. Nonetheless, their status remains weak. Moscow does not consider the Arctic to have a specific status determined by the presence of indigenous peoples, who are largely in the minority and have been acculturated to an ethnic Russian population. The lifestyle of the region is mostly urban, not "traditional"; and the stakes are economic and strategic, thereby falling within Moscow's remit. Moreover, Russia's reading of the Arctic is still very much shaped by the imperial past and memory of an easy conquest of territories deemed "unpopulated" (Laruelle 2014). Indigenous people moving to urban centers thus face many challenges, including competition with labor migrants coming from greater distances but who are more accustomed to urban culture.

The Arctic environment, by definition, encourages a multi-local lifestyle: people spend several years or decades of their life there and leave at retirement age; migrants travel there for work yet return home for weddings and funerals. Northerners spend holidays, often funded by their employers, with their family in the southern regions of the country; young people leave the Arctic to study in higher education institutions and come back for work.

Time and space are therefore complex and fragmented, with individual and family lives unfolding over different locations and chronologies. This further accentuates the feeling of insularity among Arctic inhabitants, who refer to the rest of the country through revealing terms such as "Earth" (*zemlia*), "the mainland" (*materik*), or "great or central land" (*bol'shaia or sredniaia polosa*). But it also creates new imaginaries and a mental atlas of Russia and Eurasia: Norilsk inhabitants tend to feel closer to St. Petersburg; Murmansk residents visit Tromsø or Kirkenes, at the Norwegian border, more often than they go to Moscow; inhabitants of Sakha-Yakutiya and the BAM regions may access China more easily than Russia's European regions and drive used Japanese cars with the steering wheel on the right. Arctic cities with important migration flows from Azerbaijan, the North Caucasus, or Central Asia offer more flights to Makhachkala, Baku, and Dushanbe than to many Russian cities, and so on.

Through urbanization, the Arctic region as a whole, and Russia's Arctic in particular, is experiencing drastic cultural changes that remain largely understudied. This edited volume fills the gap by providing the first study of these

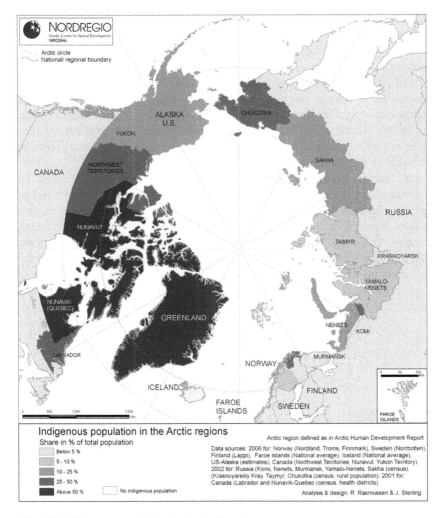

Map I.2 Indigenous population in the Arctic regions.

new mobility patterns and the resulting birth of new urban Arctic identities. It is comprised of original research, each chapter being based on specific fieldwork. We cover the main Arctic regions of Russia: Murmansk, the oil and gas fields of Western Siberia, the Norilsk region, Yakutsk and southern Yakutiya, and the BAM railway line in the Far East. Some of them, such as Western Siberia, southern Yakutiya, and the BAM region are not, strictly speaking, Arctic regions in the sense that they are located largely below the Arctic Circle. However, in terms of permafrost cover, extreme climate conditions, and remoteness, they are considered as offering quasi-Arctic situations. This is reinforced and validated by the administrative division of Russia's territory, which recognized some regions as "Arctic" or "Far North" (*krainii sever*), and some others, more southern, as "equivalent" to the former ones, with both being entitled to several financial incentives (Laruelle 2014) as seen in Map I.3.

The volume is divided into three main sections. The first one investigates the current demographic, political, and economic context surrounding new patterns of Arctic mobility. Contemporary Russia is a highly centralized country, challenging to manage due to its spatial immensity, and far removed from the federal experience of, for instance, Canada. Central authorities and regional powers are in a state of constant negotiation, formal and informal, of their relationships, in terms of political loyalty, financial dependency, and economic decision-making autonomy.

This section begins with a chapter by Nikolay Petrov that overviews the so-called Asian North, comprising the Urals, Siberian, and Far Eastern federal districts, and their demographic and political situation. Russia's politics tends to be seen from a Moscow-centered perspective and overlooks important regional diversity. Arctic regions offer a range of political climates, with some relatively open and competitive – in the Russian context of growing authoritarianism – and others opaque and monolithic. This local politics element is key to help us move from Moscow-centered Arctic policy to everyday realities on the ground, which are specific to each region and share few common features.

■ Regions of Far North

▨ Regions equated to Far North

Map I.3 The Russian Far North and its administrative status.

Chapter 2, by Genevieve Parente, pushes further away the notion of "Arctic politics" and investigates the role of the Union of Arctic Cities in representing the weight of Arctic cities and their political personnel in lobbying for their interests at the federal level and within the presidential administration. Murmansk and Norilsk, as the two biggest cities above the Arctic Circle, play a critical role in representing and defending Arctic political but also business interests in Moscow. Chapter 3, by Aimar Ventsel, examines the role of the market economy and the rise of a services economy in the capital city of Sakha-Yakutiya. He shows the trajectory of many "reluctant" businessmen who see their involvement in the private sector more as a way to cope with socioeconomic changes than as an embrace of entrepreneurship. Small businesses, often in the retail sector, function in a freer environment based on personal relationships, and assiduously avoid any relations with state structures.

The second section of the book explores three mobility patterns in Russia's Arctic regions: mobility through social capital, mobility through long-distance commuting, and mobility for indigenous people whose traditional ways of living have been profoundly affected by changes in the natural landscape (the disappearance of hunting and fishing spaces and species) and in the communities' everyday life and world representations. In all three cases, moving to the Arctic for work promotes upward social mobility. Working in a harsh environment and in often challenging – for the body as well as for family relations – conditions helps consolidate individual and family upward trajectory, not only because the salaries offered in the North are generally higher than in the other regions of the country, but also because working in the Arctic is interpreted as a sign of self-realization and evidence of a strong work ethic. The Arctic as an upward mobility mechanism is not specific to Russia: it can be found in other circumpolar regions and, in other contexts, for instance in the "far west" move of American society in the nineteenth century or in Alberta in today's Canada.

In their chapter, Nadezhda Zamyatina and Aleksey Yashunsky analyze the mobility of young Northerners, mostly from Norilsk, Dudinka, and Igarka. They demonstrate the presence of a multigenerational family cycle South–North–South based on regular relocations between the Arctic place of work and the city of origin of the family. However, they also show that migration to the Arctic fosters social upward mobility, allowing families, after decades in the North, to relocate not to the village or small town of origin, but to the big regional capital. Their research, last but not least, investigates the critical issue of "mental proximity" and the cachet that some cities such as St. Petersburg display among Northerners, an attraction that is partly but not exclusively explainable by the prestige of some higher education institutions in training those destined to work in Arctic conditions.

In Chapter 5, Gertrude Saxinger tackles the puzzle of long-distance commuting (LDC) for oil and gas industry workers in Western Siberia. This mode of mobility, which has been less common in Russia than it is in Canada or Australia to conquer new spaces and their resources, is becoming more widespread in the current Russian context of both growing state control over resource extraction and neoliberal practices that go against the Soviet welfare experience. LDC helps firms to bridge the current shortage of highly qualified personnel in the local labor market through creating an interregional pool of workers. It also allows people from socioeconomically disadvantaged regions in central and southern Russia to escape unemployment and to pursue upward social mobility. For LDC workers, mobility can appear as both a curse and a blessing, creating life divided between two locations, two sociability networks, and pushing individual and family identities to adapt to these multi-locality patterns.

The last chapter of this section, by Vera Kuklina and Natalia Krasnoshtanova, analyzes the growing urbanization trends among indigenous peoples. In Soviet times, urban indigenous were part of the local intelligentsia, promoted through the *korenizatsiya* (indigenization) policy. Yet individuals who receive higher or specialized education are less likely to return to traditional ways of life: they cannot find suitable jobs and feel uncomfortable with the conditions offered to them. Even those continuing to live "traditionally" have higher expectations in terms of everyday comforts, household appliances, mobile communications, and motorized transports. This dilemma of preserving indigenous cultures in contemporary conditions is not specific to Russia and has been studied in detail for many other countries. Since the collapse of the Soviet Union, a new trend of urbanization among indigenous groups is noticeable: towns that host an industry – mostly extraction or construction – or have regional administrative status attract individuals from small settlements, as they offer higher standards of living, fewer shortages of goods, and more jobs. Indigenous families often try to connect at least one of their members to these economic sectors in order to settle in towns, where they can also sell traditional products from hunting, fishing, or herbal medicine at a higher price. Both the Baikal-Amur Railroad (BAM), dating from Soviet times, and the more recent Eastern Siberia–Pacific Ocean oil pipeline (ESPO pipeline) form new "highways" not just for transporting goods, but also for urbanizing the indigenous population.

The third section of the book deals with another consequence of the increasing urbanization of the Arctic and of new mobilities: the birth of a more multicultural urban environment. Russia's Arctic and sub-Arctic cities have always displayed a certain level of multi-ethnicity, reflecting the national diversity of the Soviet Union. People from all the republics, mostly the Slavic ones, were engaged on these pioneer fronts to build Soviet infrastructure and embody socialism. Yet, all shared a common Soviet culture, marked by a common language – Russian – and uniform, Soviet ways of life. Over the last 25 years, diversity has increased. Once they became independent, the former Soviet republics dissociated from the Soviet mold, and a new generation of post-Soviet citizens emerged with more polarized ways of life, cultural practices, and memories of the common past. This gave rise to a more developed urban multiculturalism or at least an increased cultural differentiation at the same locality, with old generations of urbanites and Northerners having to coexist with newcomers. While Soviet-era urbanization was mostly carried out by Russian or Slavic people, in the post-Soviet era urban migration has been driven partly by labor migrants. Many young people travel to the Arctic to find work on extraction sites and industrial firms, often coming from regions in deep socioeconomic crisis, such as the North Caucasus and Central Asia.

In Chapter 7, Olga Povoroznyuk explores the growing multicultural context of medium-scale cities (i.e., Ust'-Kut, Severobaikal'sk, Tynda) and more typical towns (i.e., Novaya Chara) along the BAM railway. There, indigenous Evenki and Russian Old Settlers live side-by-side with migrants of different ethnic backgrounds. Post-Soviet social dynamics such as a growing reliance on extractive industries, the end of the Soviet welfare state model, the race to access state subsidies, and job competition in the private sector accentuate the different social and cultural identities of each group of urbanites. Indigenous people (*aborigeny*), BAM builders (*bamovtsy*), industrial shift workers (*vakhtoviki*), and more recent labor migrants (*priezzhie*) compete for a new urban legitimacy.

In the next chapter, Marlene Laruelle, Sophie Hohmann, and Alexandra Burtseva look at the case of Murmansk and its population mobility patterns. The most populated city above the Arctic Circle, with about 300,000 inhabitants, Murmansk offers a typical, Soviet urban design but unlike most of Russia's Arctic cities it benefits from economic diversity. Murmansk's prosperity is based not on extractive industries but on its commercial and military port activities, its status as a regional capital, and its close interaction with its Scandinavian neighbors. The city welcomes a growing ethnic diversity with many migrants from Azerbaijan and Central Asia who find new work opportunities while also internalizing the spirit of the pioneer front and their newly acquired identity as Northerners.

A similar phenomenon has been observed by Denis Sokolov, whose chapter concludes the volume, in his research in the oil and gas cities of Western Siberia. Examining the massive labor migration from the North Caucasus and especially from Dagestan into the Tyumen region, he investigates the creation of a new, dual identity for these migrants: that of being both "from the South" – facing discrimination and xenophobia by the local population, and strengthening a religious, Muslim identity – and "from the North" for their own communities of origin, where the brand of being a Northerner is a sign of upward mobility and personal success. In all the cases discussed here, one of the main findings is that newcomers, whoever they are, rapidly internalize their new Arctic identity, a topic that merits further research.

This collaborative research project was made possible thanks to a grant from the Norwegian Research Council devoted to the study of Russia's Arctic urban sustainability, received by the Barents Center at Tromsø University and the George Washington University for 2013–2015. Historically, the majority of studies on the Arctic tended to focus on what remains "traditional," with the underlying assumption that it needs to be preserved. Today, a new body of literature has emerged that, on the contrary, explores what is *changing* in the Arctic, without offering a value judgment on these changes, but instead with the goal of increasingly de-specifying Arctic studies and linking them to current research on globalization processes. With rapid changes occurring in the Arctic, the study of the region is becoming a critical prism through which scholars – as well as policymakers – can look at current world evolutions that share the common theme of growing urbanization, with its associated challenges and opportunities in terms of mobility patterns and environmental and social sustainability.

Part I

An evolving demographic, political, and economic context

1 Depopulation of Russia's Asian North and local political development[1]

Nikolay Petrov

The Northern section of the Asian part of Russia occupies 7.2 billion square kilometers, almost half of Russia's enormous territory, but it has just 2.2 million inhabitants. That amounts to just 1.5 percent of Russia's population, and its share has been decreasing. Russia's Asian North, as of late 2015, is comprised of six subjects of the Russian Federation: (1) Yamal-Nenets autonomous district, (2) the Northern part of Krasnoyarsk krai, (3) Republic of Sakha (Yakutiya), (4) Magadan oblast, (5) Chukotka autonomous district, and (6) Kamchatka krai, as well as the Norilsk industrial district. Administratively, this territory consisted of nine separate regions prior to 2007, when the Taimyr (Dolgan-Nenets) and Evenki autonomous districts became part of Krasnoyarsk krai and the Koryak autonomous district merged with Kamchatka oblast. A large part of these territories is located beyond the Polar Circle and belongs to the Arctic.

These vast areas are sparsely populated, and the settlements tend to be very isolated. They look like islands in a vast empty space, and, indeed, the term "mainland" (*materik*) is often used to refer to the rest of the country. Intensive colonization of the Asian North began in the 1930s when Stalin's labor camps – the "GULAG archipelago" – were built in Magadan/Kolyma, Taimyr/Norilsk, and other places. The earliest industrial development of these areas was conducted by prisoners, and later by "organized labor migrants" attracted by high salaries and other benefits. Through the 1970s, Soviet industrial development led to enormous population increases but also to serious shifts in the ethnic composition due to the inflow of Russians, Ukrainians, Byelorussians, Azeri, Tartars, and other labor migrants.

The share of the remaining indigenous populations, which in many cases gave their names and administrative designations to many of the Northern regions, is very low. The indigenous ethnic groups include peoples of the North, both large groups, such as Yakuts, and smaller ones, such as Nenets, Dolgans, and Koryaks. There are also descendants of ethnic Russian migrants from the seventeenth and eighteenth centuries, known as Russko-ust'yintsy,[2] Yakutyans,[3] Pokhodchans,[4] and Markovtsy.

Since the final Soviet census, taken in 1989, the population of Russia's Asian North has dropped by a quarter – from 2.945 million to 2.245 million in 2015, while the population of Russia as a whole remained almost unchanged (147.4 million in 1989 versus 144 million – without Crimea – in 2015). Over the past 25 years, and especially in the 1990s, the Asian North lost a considerable part of its population – nearly one-fourth of its Soviet-era inhabitants. Yamal-Nenets autonomous district, known for its large-scale gas extraction, appears to be the only Asian North region that has registered a slight population gain, while all others experienced population drops. In Chukotka, for example, the population is one-third of its Soviet-boom peak. More recently the population outflow has slowed down, and a few regions have even seen a slight growth. The transition period is probably over and the settlement pattern has adjusted to the new, post-Soviet economic and social realities.

The depopulation of the North is the result of the mass out-migration to the "mainland" of Russians, Ukrainians, and Byelorussians, workers who had been brought here earlier for the industrial development of the North. Their exodus has led to an "ethnic shift" – a visible increase in the share of the indigenous Northern peoples.

In terms of political development, the significant migration outflow has had multifaceted effects. On the one hand, the most active segment of the population, from a socioeconomic point of view, is the one that is leaving. On the other hand, those who stay, especially in the severe conditions of the Extreme North, feel more closely attached to the territory and are more likely to exhibit local and regional patriotism. The out-migration has also led the remaining population to concentrate in a smaller number of centers, which may ameliorate the sense of fragmentation inevitably arising in sparsely inhabited regions. Traditionally, dispersed settlements separated by long distances result in a highly centralized, quasi-Soviet almost feudal political structure.

A number of important administrative changes have taken place since the collapse of the USSR. Of the six cases of regional reconfiguration that occurred in Russia in the mid-2000s, three were in the Asian North; the number of the federation subjects there was reduced from nine to six. There were objective reasons for this development. First, a majority of the autonomous districts that proclaimed their sovereignty as part of the early 1990s "parade of sovereignties" are located in the North. Yamal-Nenets and Chukotka, for example, unilaterally upgraded their administrative status. Second, their ability to be self-sufficient has declined due to population losses and changes in their economic structures. However, the Yamal-Nenets district is being increasingly absorbed by its "mother" Tyumen region, and Chukotka has been in free drift since 1992.

In this chapter I evaluate the political impact of the population outflow from Russia's Asian North, and the level of democracy development in each of the local administrative entities. For that I use data on regional socioeconomic and political monitoring compiled by the Carnegie Moscow Center from the mid-1990s through 2012. I also use additional findings made under the aegis of the Higher School of Economics project on analysis of democratic institutions development at the national and subnational levels.

Democracy development at the national and subnational levels

In order to understand properly the multidimensional process of democracy development in large and heterogeneous countries, such as Russia, one should analyze not only the national, but the regional level as well. Not only can democratization at the regional level differ essentially from the national one, but the two levels can even, at least for a while, develop in opposite directions. This can be the result of mere inertia when the shifts at the national level are ahead of those at the regional one – whether in the direction of democratization, as was the case under President Boris Yeltsin in the 1990s, or in the direction of de-democratization or authoritarianism, as has been the case under President Vladimir Putin during the last decade. Otherwise, the side effects of the shifts at the national level may push the regional development in the opposite direction. The latter case can be illustrated by the Kremlin's effective efforts to weaken regional governors as national actors in the early 2000s. Despite the weakening effect on democracy at the national level, these moves led to increased political competition at the regional level, where governors toiling under the shadow of the "father of the nation" could no longer play the role of "fathers" of their regions (Petrov 2000).

Back in 2005, we wrote that "political reforms started by the Kremlin and, first of all, the abolishment of direct gubernatorial elections had a short-term positive effect at early stages followed by a negative one in the longer run. At early stages legislatures started to play a more independent role" (Petrov 2005). Indeed, the competition among regional political elites grew more intense, and longstanding connections between political and economic elites were broken. However, this effect proved to be the calm before the storm. The 2005 shift from elected to appointed regional heads struck a blow not only against elections as such, but also on the feedback between authorities and society, on political pluralism, publicity, and openness of political life. These negative effects will be fully felt in the future (Petrov 2005).

The problem of how democracy in different regions corresponds to that at the national level is far from trivial. Regional and national trajectories look like a tube or a coaxial cable with the national trajectory not necessarily at the center: at some point the national trajectory can run higher than regional ones, at another point it can run lower. In order to describe and analyze democracy in regions in comparison to pluralism the national level, one should use national ratings, such as those by Freedom House, and compare them at different points in time.

Lack of regional democracy scores

The problem, however, is that while national-level democracy ratings are available, similar estimates of democracy in the regions are not. Those made by the Carnegie Moscow Center in 1998–2013 present (see Table 1.1) a comparison across regions for the same time period, rather than the dynamics over time for any particular region (Petrov and Titkov 2013; Petrov 2004).

Table 1.1 Main political trends in Russia's Asian North regions (the lower the number, the more authoritarian; the higher, the more competitive)

Administrative region	Contestation		Responsiveness		Political design/Division of powers		Resulting rating
	Compeitveness in elections	Opposition capabilities in legislature	Feedback in elections	Administrative rotation	Legislature's autonomy	Local selfadministration' autonomy	
14 Sakha (Yakutiya)	2.8	2	3.0	3	3	4	24.9
26 Krasnoyarsk krai	2.5	2	4.1	3	4	2	24.0
24 Kamchatka krai	3.0	1	3.6	3	4	2	19.9
81 Khanty-Mansi district	2.5	1	4.1	3	3	2	15.6
82 Chukotka district	1.0	2	2.9	2	4	4	14.6
83 Yamal-Nenets district	1.8	1	2.0	3	4	2	10.3
51 Magadan oblast	2.3	1	4.0	2	2	2	9.8

Source: Nikolay Petrov.

There are two ways to evaluate the width of a regional corridor – how much a region oscillates on its own and in comparison to the national trajectory. One way is to build trajectories for a number of regional cases that would represent the whole picture as fully as possible. The other way is to build several vertical snapshots for the whole set of regions on the basis of measurable indicators, such as behavior in consecutive electoral cycles.

Several hypotheses explaining variations in democracy development in the regions should be tested. Possible explanatory variables include the size of the region; the geographic position in terms of North–South (see Figure 1.1) and center–periphery; the urban–rural divide; the economic and spatial structure, which can either promote or suppress political competition; institutions and path dependence; political culture and traditions; and leaders and political elites. It is perhaps political culture that can explain the model of mosaics, with the Urals and Russia's Northwest being more democratic than other macro-regions.

The initial impetus model works here as well, as the April 1993 elections of regional heads in eight regions illustrate. Elections were held in regions where Yeltsin's 1991–1992 reformist appointees faced serious challenges from more con-servative regional legislators. Five regions out of eight – Smolensk, Bryansk, Orel, Lipetsk, and Penza – belong to the so-called red belt surrounding Moscow from the west and south, while three others – Chelyabinsk, Krasnoyarsk, and Amur – are located far east of Moscow. The incumbents lost to former conservative heads of regional administrations in all but one case (Krasnoyarsk). However, almost none of the 1993 winners managed to keep his office in the next gubernatorial election cycle, which took place in 1996 in Amur, Bryansk, and Chelyabinsk and 1998 or later in the rest of the regions. The 1993 initial gubernatorial elections thus launched the mechanism of regular democratic replacement of regional leaders through elections.

Latitudinal distribution of population in Russia (people per degrees north)

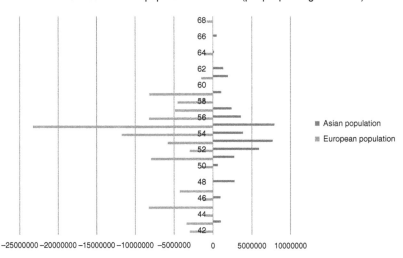

Figure 1.1 Distribution of population of Russia in European and Asian parts by latitudes.

Source: Nikolay Petrov.

Moscow and St. Petersburg, designated as the only two national capital cities, constitute a very special case. Their behavior varies during different phases of the country's democratic development. On the one hand, at a time of stable national development, their democratic development was close to the average regional level. On the other hand, they led the democratization process at a time of turbulence, both in the late 1980s–early 1990s and during the 2011–2012 political protests over disputed election results.

The national government can either promote or suppress regional democracy. Our previous studies show that regardless of the center's intentions, its direct interference into regional affairs – usually in the form of personnel replacements – has a negative effect. Specifically, they diminish democracy in the regions and stunt their internal political evolution (Petrov and McFaul 1998, 156).

The relation between the general level of democracy in the regions and the 2011–2012 anti-Putin protests should be considered in order to make additions and corrections to democracy ratings if needed, as well as to explain the correlation between democracy and the protests' spatial patterns. So far, it looks like the protest patterns can be explained by the hierarchical diffusion of the socioeconomic innovation patterns, rather than by regional differentiations, with Moscow leading, followed by St. Petersburg, and other large population centers still farther behind.

Regional portraits[5]

Regional variation in democracy is presented in Table 1.1 and in Figure 1.2. Russia's Asian North regions can serve as a good case study to test various hypotheses explaining the dynamics of democracy at the regional level, and give us an insight into local political development in the Arctic.

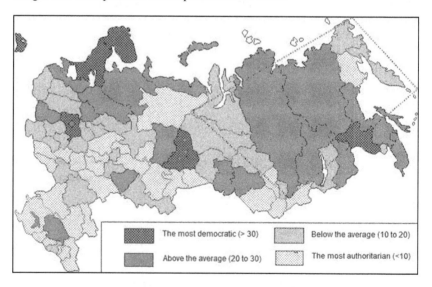

Figure 1.2 Democracy ratings by region.

Source: Nikolay Petrov.

Yamal-Nenets autonomous district (YaNAO)

Yamal-Nenets has undergone a wholesale population exchange. Thanks to the influx of migrant workers in the Soviet era, newcomers far outnumber the locally born. Only one in every four of its inhabitants was born in the greater, Soviet-era Tyumen. The district has experienced several population explosions: its population grew sixfold in the 1970s and 1980s, and tripled again in the post-Soviet era. The district has experienced not only the high natural population increase typical of newly developing regions with relatively young populations, but a migration inflow as well, unlike other regions of the North.

Ethnic Russians comprise more than 60 percent of the population of YaNAO, and their share continues to grow. Indigenous people account for roughly 10 percent, including 5.9 percent of Nenets and 1.9 percent of Khanty, and their share is growing as well, while the share of ethnic Ukrainians and Byelorussians, which was close to 20 percent at the time of the USSR's disintegration, has dropped almost twofold. The population of the district capital, Salekhard, which was the region's only town in the early 1970s, has grown to 48,000 – up from 22,000 some 40 years earlier – insignificantly by local standards, and it is currently outnumbered by younger towns, such as Novy Urengoy (116,000[6]) and Noyabr'sk (107,000).

Generally speaking, YaNAO is a Gazprom company district. The energy giant's monopoly there has been slightly impinged upon by oilmen (Rosneft) and independent gas producers like Novatek, Itera, and Northgas, which exploit former Gazprom subsidiaries. Political life in the district is not very active and is mostly confined to individual towns rather than in the district as a whole. The party system is more imagined than a reality. New district heads have appeared only twice in 25 years: in 1994, Lev Bayandin, the incumbent since Soviet times, was replaced a by local resident, Yuri Neyolov, who had previously spent about a decade in high positions in Tyumen. In 2010 Neyolov was replaced by Dmitry Kobylkin, an engineer-geologist and former top manager of a Novatek subsidiary, who had worked as a subdistrict head in one of the okrug's seven *rayons*.

According to democracy ratings, YaNAO in the 1990s ranked in the middle of the regions, with results close to the nation's average. The voting pattern in federal elections looked more democratic due to higher vote levels for the "Our Home Russia" party, which benefitted more from Gazprom's administrative and corporate support rather than the democratic aspirations of the electorate. In the 2000s, when "sovereign democracy"[7] flourished, Yamal became an "electoral anomaly" due to tight administrative control in rural areas and strict corporate discipline in the oil and gas towns. Both the official turnout and the results of the "party of power" (United Russia) were extremely high here; election experts believe that the YaNAO district experienced massive fraud. Local self-administration is extremely weakened, the district is among the first to eliminate direct mayoral elections, nongovernment media is absent, and political parties are mere imitations. There is a higher degree of political competition in Noyabr'sk, where both oil and gas producers are present.

Taimyr (Dolgan-Nenets) autonomous district

Located beyond the Arctic Circle, Taimyr is a giant, almost undeveloped area surrounding the industrial city of Norilsk. It lost one-third of its population in the 1990s and 2000s. In 2007, when it merged with Krasnoyarsk krai, there were 38,000 inhabitants there, two-thirds of whom – 25,000 – lived in Dudinka, a port on the Yenisei River and Norilsk's satellite. Out-migration has caused a sharp decrease in the share of ethnic Russians, now about one-half, and an increase of the indigenous peoples' share – Dolgans now comprise 14 percent and Nenets nearly 8 percent.

The district had three governors in 15 years, starting with Gennady Nedelin (1991–2001), who was born in the southern part of Krasnoyarsk krai and worked for about two decades as a deputy district head before being appointed district head. Next, Muscovite and Norilsk Nickel CEO Alexander Khloponin won the post through an election (2001–2002). He was succeeded by the former mayor of Norilsk, Oleg Budargin (2003–2007), who had been a top manager at Norilsk Nickel.

When it merged with Krasnoyarsk krai, Taimyr became part of a huge municipal rayon, and direct elections were abolished. It is formally headed by the speaker of the rayon council; the rayon administration is headed by a contract employee. The district is restricted; Russians must have a special permit, issued by law enforcement, to travel to the area. Democracy ratings put Taimyr district in the middle of the list due to the externally imposed democracy. Moscow wants to control the region, due to its proximity to Norilsk, a large company town for a strategically important industry.

Evenki autonomous district

Evenki's territory is huge, but its population is tiny: just 17,000 residents in 2007. There are almost no roads in the district; instead, people travel by *zimniks*, the icy "winter roads" on frozen rivers. Hunting and fishing are major occupations of the local inhabitants, most of whom are either unemployed, retired, or rely on salaries paid out of the government budget. There are practically no political parties or local media.

Before its merger with Krasnoyarsk krai, Evenki was the smallest subject of the Russian Federation in terms of the number of inhabitants. It has lost more than one-quarter of its population since 1989, causing a shift in the relative ethnic distribution: the share of ethnic Russians dropped to 62 percent, while the Evenki increased to 21.6 percent.

Of the three governors in the district since 1991, two have been representatives of the local Communist *nomenklatura*: Anatoly Yakimov (1991–1997) and Alexander Bokovikov (1997–2001). The third was a top manager from the oil company Yukos, a Muscovite, Boris Zolotarev (2001–2007).

Evenki ranked among the bottom ten regions of Russia in terms of democracy ratings.

Norilsk industrial district

This urban complex is centered around the massive Norilsk mining and metallurgical plant. The district is comprised of Norilsk itself and two satellite towns – Talnakh and Kayerkan. Construction of the plant began in 1935 and was carried out by GULAG prisoners. The population steadily increased through the mid-1980s, peaked at 250,000, then started to decline. Currently, it is about 176,600. Norilsk is not only the world's second largest town located beyond the Arctic Circle after Murmansk, but it is also the fourth most windy one and a major source of air pollution.

Norilsk has a specific political culture that combines elements of a "Northern brotherhood" mentality and "Northern patriotism" with developed protest sentiments and institutional nihilism. With elections that are presumed to be rigged, the region is infamous for its extremely low turnout on election days. When the "against all" ballot option existed, that tally would overwhelmingly win.

The Norilsk population mostly consists of workers who arrived in the second half of the twentieth century and their children. There are also descendants of GULAG prisoners who were released in the mid-1950s and could not return to their pre-GULAG hometowns. Historically, the town was administered by the Norilsk plant management. From 1990 to 1995, its mayor, Vasily Tkachev, a former truck driver and head of a construction enterprise, argued that more tax revenues should be retained in the city, instead of being sent to Moscow, and started a petition asking President Yeltsin to subordinate Norilsk to Moscow instead of Krasnoyarsk. He was promptly sacked by the Krasnoyarsk governor, but managed to defeat the acting mayor in the 1996 elections and remained in this office until 1999. That year he was convicted of bribery and replaced by a top Norilsk Nickel manager, Oleg Budargin, who subsequently won the mayoral election of 2000.

Budargin left to head Taimyr district in 2003, triggering a new mayoral election. In April 2003 the first round ended with the victory of a rebellious trade unions leader, Valery Melnikov, who was later disqualified by the election commission. A new election again produced a Melnikov victory, despite the local administration's efforts. Melnikov won a seat in the State Duma in 2007. Instead of organizing another direct mayoral election, the town council decided that Norilsk would switch to a model in which the city would be governed by the council speaker and a hired city manager.

Norilsk has not been included in democracy ratings because it never was an independent subject. It looks, however, that Norilsk is more democratic than Taimyr and almost as democratic as Krasnoyarsk krai as a whole, despite being a closed company town. The city's moderate depopulation has not had any significant impact on its democratic development.

Krasnoyarsk krai

The population of Krasnoyarsk krai grew quickly after World War II, reaching 3 million in 1989. It subsequently declined until 2010 and dropped down to 2.828 million before rebounding to 2.859 in 2015. Four-fifths of the krai's population, including Krasnoyarsk's 1.052 million inhabitants, lives south of the Angara River – on one-tenth of the krai's territory. The local population is deeply rooted on this territory; more than 85 percent are second- and third-generation inhabitants, the descendants of those who came here with waves of the Russian colonization between the seventeenth and nineteenth centuries; more than 90 percent are hereditary Siberians.

Political life is lively both at the elite and citizen levels. The party system is rather competitive, so no local governor has ever enjoyed absolute power. There have been six governors in 24 years. The first to come was Arkady Veprev, USSR peoples' deputy, the head of a famous Soviet collective farm. He failed to establish good relations with regional elites and quit in 1993, leaving behind his deputy, economist Valery Zubov, whose career began at the university and continued at the new Krasnoyarsk stock exchange. He was reelected in 1993 – the only one of the seven incumbents to be returned by the voters. Zubov headed the region until 1998 when he lost the election to General Alexander Lebed, a charismatic federal-level politician who had come in third in the 1996 presidential race.

The four years under Governor-General Lebed were marked by numerous personnel replacements and elite infighting. Lebed's tenure ended abruptly when he was killed in an air crash in 2002. The subsequent special election was won by an "outsider" – Taimyr governor and former Norilsk Nickel CEO Alexander Khloponin, who received 48 percent of the vote in the second round against Krasnoyarsk native Alexander Uss, who had the backing of regional elites. The regional election commission twice invalidated the election results, leading President Putin to intervene and appoint Khloponin himself. Khloponin served happily for eight years, leaving in 2010 when he was appointed to the high-ranking cabinet position of deputy prime minister. He was replaced by Lev Kuznetsov, a Muscovite who had been his deputy, first at Norilsk Nickel and then in the Taimyr and Krasnoyarsk administrations. In 2014 Kuznetsov succeeded Khloponin as a cabinet member in charge of the Northern Caucasus. After that, Viktor Tolokonsky, a former governor of Novosibirsk oblast and later the presidential envoy to Siberia, was appointed governor of Krasnoyarsk.

In terms of its political elites, Krasnoyarsk plays a special role. First, it attracts the attention of federal elites, including General Lebed and the managers of Rossiiskii Kredit Bank, who backed him. Second, this region has sent the largest number of representatives (after St. Petersburg) to Moscow, including Deputy Prime Ministers Alexander Khloponin and Olga Golodets, as well as cabinet ministers Alexander Novak and Lev Kuznetsov, among others. All of them, however, started their careers at Norilsk Nickel and can be regarded as business managers who moved first into regional politics and later into federal politics. The Norilsk Nickel team circulated from Moscow to Norilsk, then to Taimyr, Krasnoyarsk, and back to Moscow. This kind of deployment to a region and a subsequent return to "headquarters" follows the general cycle of "regionalization–centralization."

Krasnoyarsk krai traditionally receives top scores in the democracy rating. Its electoral rating is high as well although it has fluctuated over time: 7.6 (2001), 12.6 (2004), 9.7 (2009), 8.9 (2012), 10.3 (2014), on a scale of 3 to 15. Regionalism is strong in Krasnoyarsk; in the past, it was manifested by solid votes for regional parties and blocs. After local parties were banned, that vote went to regional franchises of federal parties in which local groups kept substantial independence.

Sakha-Yakutiya Republic

The population of Sakha-Yakutiya has been decreasing slowly since 1990s, and in a matter of 25 years has decreased by 10 percent. There is an outflow of Russians, Ukrainians, and other migrants (as well as their children) who arrived 30–40 years ago to develop mineral resources. The relative share of the indigenous population – Yakuts, Evens, Evenkis, etc. – increased by 50 percent in the 2000s and continues to grow. It is equally high in the city of Yakutsk, whose population keeps growing rapidly (up 20 percent between the 2002 and 2010 censuses) due to both an inflow from Yakutiya's rural regions and labor migrants from Central Asia and the Caucasus. In 1990, a little more than one-sixth of the population was concentrated in the city of Yakutsk; today almost one-third of this region's population – 294,000 out of 957,000 – lives in Yakutsk.

Although regional leaders (governors) hold their offices for many years, local political life is characterized by intensive competition and pluralism with a complicated balance of different "clans," both territorial and corporate.

Three leaders have governed Yakutiya during the past 25 years. The first president of the Yakutiya Republic, elected in 1991, was an ethnic Yakut, Mikhail Nikolayev, who made his career in the Komsomol and the Communist Party during the Soviet period. In 2002 he was replaced with his vice president, Vyacheslav Shtyrov. A local Russian, Shtyrov became head of the Yakutiya jewel extraction company "Diamonds of Russia – Sakha" in 1995. Shtyrov stepped down in 2010 and was replaced by an ethnic Yakut, Yegor Borisov, who had been the head of cabinet under Shtyrov.

Yakutiya ranks from the low 50s to high 60s in the expert democracy rating. Its electoral rating at the end of the 1990s was among the highest in Russia, then it declined until 2009 when it fell below the nation's average. In recent years, however, the region has rebounded and its ratings have been higher than Russia's average due to its greater political competition. All major federal parties are fairly active in Yakutiya, and the 2014 election for the regional head was among the most competitive in Russia.

Chukotka autonomous district

The population of the district shrank in the 1990s more than threefold – from 164,000 in 1989 to 54,000 in 2002 – and has stabilized in recent years at 50,000. In previous decades, the district was a net recipient of migration, with relatively few residents who were born there and a high inflow. According to the 2014 census, ethnic Russians today make up less than 50 percent of the local population, while the share of Chukchi has increased to more than 25 percent. Taken together with the Evens, Evenki, and others, the indigenous peoples comprise about one-third of the population. The district's center, Anadyr, contains a little more than one-quarter of the population with the majority of the Chukchi living outside urban settlements and maintaining a traditional lifestyle.

Political parties are almost non-existent, and leading candidates show suspiciously high vote counts in elections. Chukotka has had three governors in 25 years. The first, Alexander Nazarov (1991–2000), was a member of the Soviet *nomenklatura*. In 2000 he voluntarily chose not to run for reelection, paving the way for oligarch Roman Abramovich (2000–2008), then a State Duma deputy from the district, to assume the office. Abramovich not only started to pay his income tax in the district, but he transferred some of his businesses to the district and sponsored several local infrastructural and cultural projects. Although he expressed his desire to leave several times, Putin insisted that it was Abramovich's "social responsibility" to provide for Chukotka's financial well-being. Ultimately he was allowed to leave only in 2008, when he was replaced by his deputy, Roman Kopin, who arrived earlier with Abramovich from European Russia.

Chukotka has long ranked near the bottom of the democracy indicators, and dropped even further after Abramovich and his team departed. It now ranks as one of the three least-democratic regions of the country. The district's electoral rating has also plummeted. Today Chukotka has the worst electoral rating in the Russian Federation.

Koryak autonomous district

Koryak was a separate subject of the Russian Federation until July 1, 2007, when it became a district of Kamchatka krai. The population is decreasing due to outflow: from 39,400 in 1989 to 22,600 in 2007 and only 17,600 in 2014. It therefore lost two-thirds of its population in just 25 years. Consequently, the relative share of the indigenous population has increased, from a quarter to about one-half. Approximately one-third are Koryaks, and the rest are other peoples of the North.

Political competition in Koryak occurs between two major interest groups: fishermen and miners. The district has experienced a high turnover in its top leadership, with four leaders in a little more than 15 years. The first governor was Sergei Lyoushkin (1991–1996), a locally born engineer-economist who came to politics in the *perestroika* era. He lost the 1996 election to Valentina Bronevich (1996–2000), an Itelmen supported by the fishermen bloc. Bronevich had been the Soviet-era district head before Lyoushkin. In 2000 she lost the election to the miners bloc representative Vladimir Loginov (2000–2005), who managed to be reelected in 2004 but was fired by President Putin after a year, ostensibly for failing to adequately prepare for heating season. Many observers, however, believe he was fired for opposing the district's merger with Kamchatka. The district's final governor was Oleg Kozhemyako (2005–2007), a rich fishing merchant who moved his business from his native Primor'ye to Kamchatka. He managed to facilitate the merger with Kamchatka by negotiating with the Communist-dominated district assembly.

Koryak's democracy rating is in the bottom third of the list, although higher than Chukotka and Evenki.

Kamchatka krai

Kamchatka's population grew steadily until the 1990s, then dropped by a quarter in the course of 11 years – from 479,000 in 1991 to 358,000 in 2002. The decrease slowed in the 1990s, and the population has stabilized at 320,000. Some 86 percent are ethnic Russians, and their share is growing slightly. Indigenous peoples constitute about 5 percent, including Koryaks (2.3 percent), Itelmens (0.8 percent), Evens (0.6 percent), Kamchadals (0.5 percent), and Chukchi (0.5 percent). Two-thirds of the population are located at the southernmost part of the peninsula in the regional center Petropavlovsk-Kamchatsky (183,000) and its satellite, Yelizovo (38,000), which has been essentially absorbed by Petropavlovsk.

The krai's once-lively political life has become less turbulent in recent years. The party system is rather competitive, and the numerous inter-elite conflicts are corporate or personal by nature. Over the past 25 years the region has had four governors, all of them leaving before their terms expired. The "fisherman" Vladimir Biryukov was the first.[8] He made his career in the Soviet times, was fired in the 1980s and expelled from the Communist party, but managed to make a comeback and become head of the region's executive committee in 1990. In 2000 he was replaced by a Communist, Mikhail Mashkovtsev, who won reelection in 2004 despite being the subject of a criminal investigation. Mashkovtsev resigned in 2007 and was replaced by Alexei Kuzmitsky from St. Petersburg, who, in turn, was replaced in 2011 by a local official, Vladimir Ilyukhin.

Kamchatka's democracy rating places it around 30th out of 83 subjects of the federation. In terms of electoral ratings, Kamchatka is one of the leaders, scoring far above the average.

Conclusion

The regional political dynamic of the Asian North regions generally follows the pattern of "a hundred flowers blooming" in the "wild" and hungry 1990s, followed by a period of declining activism. The region's fortunes are heavily linked with the rise of Russia's hydrocarbon prosperity and the country's overall transformation from a genuine federation into a unitary and highly centralized state. The Kremlin's interference in the political development of the Northern regions is quite noticeable, especially in the Far East. Nowadays one can hardly imagine a rebellion against the administration like the one that happened in the Norilsk mayoral elections in mid-1990s. Of course, today there are no active forces to oppose the administration and no elections, either.

The administrative realignment carried out in the second half of the 2000s eliminated one-third of the Asian North subjects of the Russian Federation. These changes played both a negative and a positive role: negative, because they encroached on the ethnic regions' sovereignty and minimized regional peculiarities, and positive in terms of democratic development because less-democratic regions were attached to more democratic ones.

Over the course of 25 years, governors were steadily disempowered, as were regional elites in general. Regional strongmen, part of the first wave of post-communist leaders, were removed and replaced with technocratic managers – originating from both the business world and the bureaucracy. Over time, governors were replaced more frequently, with an average time in office falling from 6.5 years in 2000 to 5.0 in 2005, 3.3 in 2010, and 3.7 in 2015. Putin has replaced all Northern regional governors at least once. Moreover, while in 2000 all governors, except for General Lebed in Krasnoyarsk krai, had been selected by regional elites, in 2015 there were only two regional leaders – in Yakutiya and Magadan – who could be considered protégés of the local elites. Two more, in YaNAO and Kamchatka, were "home-grown," albeit second-tier, bureaucrats. This means that during Putin's 2000–2008 presidential tenure not only has the governors' corps been thoroughly reshuffled, but also a new regime of reproduction of the corps of governors has been established. The political weight of governors, as well as of all other local political heavyweights, such as the mayors of regional centers, has decreased.

The influence of migration on the political development of these regions is complex and multidimensional. On the one hand, the outflow of the more politically engaged segment of the population generally leads to a decrease of activism as a whole, including political activism. On the other hand, the outflow of the less rooted part of population has made those who stay more eager to participate in public life. The motivations of the remaining indigenous population, however, can differ strongly from those of the more cosmopolitan Slavs. The electoral anomalies characteristic of the ethnic republics with high shares of the titular population can serve as a good example of a different motivation – being more focused on relations with their local bosses than on federal politics, they tend to vote the way that bosses/local elites want.

The new equilibrium reflects more than a mere change of quantitative parameters. The new "rotational workforce model" can radically change the political perceptions of those who remain in a territory. This phenomenon may be illustrated by the conformist voting pattern of the crews of the fishing ships assigned to Kamchatka, who vote for local candidates as ordered by their captains.

In terms of regional democratic development, there has been no direct effect from either the migration outflow, even on the scale of Chukotka, or the inflow as in the Yamal-Nenets district. In this respect, other factors appear to be more important, such as the economic diversification and citizens' critical mass formation, a more complex spatial structure, the formation and development of political elites, and the presence of efficient institutions, as well as Moscow's involvement in regional politics.

The migration phenomenon is also characteristic of the local elites, and migration patterns are changing. The eastward trend exemplified by General Lebed in 1997 and Roman Abramovich in 2000 has given way to a westward movement, best represented by Alexander Khloponin and his whole team in the late 2000s–early 2010s. Rotations of federal officials in the regions became widespread practice, in some cases including even the governors, who transformed from being the highest regional elites' representatives at the center to Moscow's highest representatives in the regions. Viktor Tolokonsky, the former Novosibirsk governor who was first appointed plenipotentiary presidential envoy to the Siberian federal district and later became the governor of Krasnoyarsk krai, illustrates this trend.

Notes

1 This chapter is based on the results of ongoing research project "Analysis of interconnection between democratic institutions development at national and subnational levels: the case study of Russia and its regions" at the Laboratory for Regional Development Assessment Methods, Center for Fundamental Research, National Research University, Higher School of Economics, Moscow.
2 Old-timers from the village Russkoye Ust'ye, who settled at the mouth of the Indigirka River at the beginning of seventeenth century.
3 Descendants of early Russian colonists who came to Siberia in the eighteenth century and assimilated with Yakuts.
4 A sub-ethnic group of Russians settled in Pokhodsk and Chersky in Yakutiya at the mouth of the Kolyma River; they descended from the seventeenth-century pioneers.
5 The author would like to thank Alexander Kynev and Alexei Titkov for their valuable comments and contribution to this section.
6 All population numbers are given for January 1, 2015, unless otherwise specified.
7 The Kremlin's term to emphasize Russia's right to a unique, non-Western model of democracy.
8 Biryukov was not a fisherman himself, but represented their interests.

2 Shaping Russia's new Arctic

The Union of Cities in the Arctic and the High North

Genevieve Parente

In May 2014, Russia established its official Arctic region for the first time through an Executive Order, "On the Russian Federation's Land Areas in the Arctic Zone."[1] The Executive Order sets the region's "borders and composition," in order to implement previously adopted federal economic development plans for the Arctic region (Russia 2013a). Dedicated federal funding and private investment for state development programs will flow to the new region, and federal funding alone is expected to reach almost two trillion rubles (about $45 million) by 2020 ("Proekt gosprogrammy").

The Executive Order was originally part of a draft federal bill, "On the Arctic Zone of the Russian Federation" (hereafter, "Arctic Zone"), which has not yet become law.[2] The bill remains stranded in the federal legislative process, to the dismay of Northern regional leaders who desire concrete plans and associated funding for the new region. The Executive Order is the only portion of the languishing bill to pass so far. However, as Raisa Karamzina, a United Russia deputy in the State Duma commented, "Much depends upon this [Arctic Zone] law: with its adoption, we will obtain social security for people living in the Far North — that includes pensions, housing issues, social infrastructure, and healthcare" (Sterlnikova 2013). A review of the draft legislation was apparently "one of the key objectives" for 2015 for the Expert Council on the Arctic and Antarctic under Russia's Federation Council, according to a May 2015 Federation Council press release (RAO/CIS 2015),[3] but by the end of 2015 it had still not been passed.

The Arctic Zone legislation is among a flurry of Arctic policy documents produced by the Russian federal government in the last few years. Together, they outline the Kremlin's vision for the new Arctic region and suggest that Moscow intends to maintain centralized control in the area, given its geopolitical importance and natural resource wealth. The establishment of the Arctic Zone as an administrative unit reflects and reinforces Russia's centralized political structures and processes in several ways. Even its form – an Executive Order signed by President Vladimir Putin – indicates the Kremlin's intentions to extend its strong role in regional development. Moscow exerts authority in the region quite clearly through territorial demarcation and setting policies to control the region's vast natural resource wealth and other economic sectors important for national economic and geostrategic security (Putsykina 2011; Orttung 2006, 2).

However, even though most of this recent Arctic legislation was negotiated at the highest levels of government, several Northern actors attempted to amend aspects of the proposed legislation around their own interests, as part of their longstanding efforts to expand their role in Arctic decision-making. This chapter focuses on the political strategies of the Union of Cities in the Arctic and the High North (hereafter, the "Union") and several of its members to extend their influence in the administration of the new Arctic Zone.[4] To assess the extent to which the Union and its members were successful in inserting their own interests in federal plans, this study examines the intervention of the Union in federal development plans for the region, as well as the attempts by two of its members to be included in the newly defined Arctic Zone.

As the Arctic Zone legislation was being formulated, the Union as an organization recommended changes in the provisions related to economic development and advocated for a larger Northern role in federal-level decision-making related to the Arctic. The Union is an interregional association established to advance the interests of Northerners. It is comprised of 52 cities and regions of the Russian Far North and equivalent areas.[5] In addition to the Union's collective advocacy, two of its most economically powerful members, the city of Norilsk and Murmansk oblast, successfully lobbied for their own inclusion in the Arctic territory established in the Executive Order. Norilsk hosts mining and metallurgy giant Norilsk Nickel and Murmansk oblast is strategically important for Arctic shipping and natural resource development, especially petroleum. Their success in swaying federal authorities likely stems from their economic importance and political connections compared to many of the smaller cities in the Union. In 2013, Norilsk and Murmansk both intervened as the bill began the federal approval process. In addition to advocating for their own inclusion in the Zone, they requested more regional spending by the central administration.

This chapter begins by describing the Union of Cities as an organization. It then turns to the Union's recommendations for federal development plans in the new Arctic Zone. It then discusses Russia's new Arctic Zone, as established by the Executive Order, and the political strategies used by the city of Norilsk and Murmansk oblast to achieve inclusion in the region and to expand their role in the administration of the region. The chapter concludes with a discussion of what their differing experiences in federal advocacy tell us about state, business, and regional dynamics in the Arctic.

The Union of Cities in the Arctic and the High North

The Union of Cities in the Arctic and the High North is an independent regional association, originally established "by the Soviet [government] and the administrations of 14 northern cities" in 1991 to look after the socioeconomic interests of the Northern regions of Russia (Union 2013). The Union has grown to include 52 member cities and regions in Russia's Far North. It is charged with "contribut[ing] to the stabilization of the economy of the North, the social status of Northerners, the preservation of natural resources, indigenous peoples, their culture and traditional crafts" (Union 2013).

The Union's administrative structure, funding, and position in the Russian political context indicate that the organization remains a creature of the Kremlin, despite its status as a regional association. Potential Union challenges to central power have been mitigated in its administrative structure and funding, in the Union Charter, and in other founding documents. Although the Union is comprised of Northern regions and represents their interests, it is a legal entity with an address in Moscow and an annual budget from the federal government of $5 million (Laruelle 2013).

The organization's main objective is to encourage regional economic development and to coordinate these efforts among its membership and federal ministries. Its mission includes directives to "[create] the conditions necessary for the effective cooperation between the participants of the Association [the Union] in matters of socioeconomic development [...] by combining the material and financial resources to carry out activities of common interest" (Union Charter 1993). The Union is also charged with the "coordination of [...] joint activities to stabilize the economic situation and social protection of northerners" (Union Charter 1993). The Union cooperates regularly with a range of other government ministries and organizations to achieve their goals, including "the Committees on the North of the State Duma and the Federation Council, ministries, committees and other executive structures of power, the Public Chamber of the Russian Federation, as well as with other associations" (Union 2013).[6]

The Union works on a range of issues primarily concerned with the "socio-economic development of their territories, by combining the material and financial resources to carry out activities of common interest" (Union Charter 1993). Presumably the "common interest" is set at the annual Union Congresses, where members meet and where much organizational decision-making occurs. The Union Charter notes that the Congress is central to the organization. "The supreme body of the Union, under the Charter, is the Congress. Between congresses, management of the association ... [is carried out by the Board and President]" (Union 2013). For many years, Congresses took place in the State Duma chamber, in Moscow, although more recently, they have begun to take place in Northern member cities. Since the 32nd Congress in Salekhard in 2012, all Congresses have occurred in the North.

The federal government details the centralized administrative hierarchy for the association in the Union founding documents. The Union's 1993 charter notes that the organization is managed directly by the federal administration, suggesting that most important powers remain with the central state. The Union is governed by the laws of the Russian Federation's Presidential decree, as well as decrees and orders of the Council of Ministers, the Government of the Russian Federation, and the Union Charter itself. The charter further clarifies that provisional regulations for the formation, registration, and operation of economic cooperative organizations between the federal government and local authorities, such as the Arctic Union, were approved by the Government of the Russian Federation in September of 1993 (Union Charter 1993 § 1.4).

There are several rationales for the Kremlin's creation and funding of the Union, given that the organization appears to hold little independent power. The Kremlin appears to have set up the Union to ensure a politically distinct organization it can control, enlisting the Union in federal plans for the region that involve controlling sectors of the economy important for economic and geostrategic security. These plans include federally directed resource development and funding distribution in support of federal programs. The Kremlin's methods of maintaining centralized control of Arctic economic and political affairs have taken several other forms, including laws limiting foreign investment and installing political appointees in corporate ownership positions.

The Union's structure also seems designed to co-opt and eliminate potential political challenges originating from the region. The Union's membership, for example, cuts across existing regional juridical administrative regions and disrupts their power.

The Kremlin has also formally enlisted the organization in federal efforts to combat decentralization. The Union charter notes that the organization must "support and implement … the main directions of the state regional policy aimed at strengthening the national integrity of the Russian Federation, as well as resistance to the elements of separatism, nationalism, and decentralization of state power in the Russian Federation."[7]

Union intervention in Arctic policy

Despite a growing preponderance of central influence, the Union is also developing as an entity that can advance some of its own interests and as a mechanism through which its affiliates can do likewise. In this way, the Union and its members are shaping the extent and character of the Arctic Zone and its governance. They also undercut, to some degree, the central state's influence in the Arctic.

Their recent advocacy suggests that, despite a strong Kremlin imprint on its structure, the Union and its members are attempting to assert themselves. They are using several platforms and strategies to advance their interests in key policy areas. The Union sees itself as the primary advocacy organization for the Arctic region and its inhabitants. Certain socioeconomic issues are particularly acute in the North or otherwise unique to this harsh environment. The association advocates at the federal level on behalf of its constituents. These issues are of abiding interest to the organization, including difficulties in providing affordable housing and adequate healthcare, and ensuring the delivery of goods and services in these often remote and inaccessible locales.

Strategically, the Union appeals particularly to the appropriate ministries in is advocacy. For example, the 28th Congress in Moscow took place at the Ministry of Housing and the Ministry of Regional Development. The organization looks to other ministries as well. In 2015, following its most recent housing-themed Congress in Naryan-Mar, a city that shares the housing woes prevalent in Northern cities, the Union prepared an appeal to the Federal Ministry of

Construction and Housing. The Union asked for the development of detailed technical regulations for major repairs in common property multi-apartment buildings, including recommended materials for upgrades and what documents are necessary to justify the scope of work (Chistiakova 2015).

However, the Union increasingly believes its input is marginalized in Moscow, and it is therefore utilizing new strategies to reassert itself in regional decision-making. This is clear in their 2013 white paper addressed to the Federation Council, arguing that the newly established "Federation Council Committee on Federal Structure, Regional Policy and Local Self-Government of the North" is neglecting socioeconomic concerns in the North, including economic policy, taxation, fiscal relations, pensions and social security, and a number of others (Union 2013). The Union noted particularly that despite Russia's recent focus on socioeconomic development in the Arctic, and associated legislative support, the committee neglects these issues. "Judging by the work plans, these [same] issues are not a priority for the new Committee" (Union 2013). Overall, the Union noted that the new Committee is far less responsive to their concerns than it used to be, when it was still the Committee for Northern Affairs and Indigenous Peoples. The new Committee was created from four previously existing ones, including the Committee for Northern Affairs and Indigenous Peoples, and appears to have streamlined their duties (Union 2013).

It is also possible that this change in the Committee's focus and activity may be part of broader changes in Russia's federal administration of the Arctic. There are strong rumors that Putin is seeking to establish an Arctic ministry to centralize regional decision-making. It is possible that federal reshuffling for such a ministry has already begun to leach duties and responsibilities from existing federal sub-committees and councils administering various parts of the Arctic. The establishment of a central ministry, "dedicated solely to the Arctic, large and strong enough to assume a wide variety of duties and responsibilities," was attempted and failed several times during the Soviet period, notably in the early 1930s (Gruber 1939 cited in McCannon 1998, 34).

The Union also leverages its leadership to influence federal Arctic legislation. Its management includes well-connected bureaucrats who reflect the organization's economic focus and who have long tenure in the region. Igor Shpektor was elected president of the Union in December 2009. He is also an elected member of the Public Chamber of the Russian Federation, where he works as an advocate for the interests of the North and Northerners. Most of the Union's Executive Directorate are bureaucrats who hail from some of the region's most politically connected and economically strong resource cities. As of August 2012, the Union's website showed nine members of the Union Board of Directors, including Oleg G. Kurilov, chair of Norilsk's City Council of Deputies.[8]

In 2013 Norilsk's Kurilov became the Union vice-president, elected by the Union delegates at the 33rd Congress held in Norilsk. Kurilov already had experience in the Union but, most important, hails from one of the Union's more powerful member cities. The association will likely benefit from elevating Kurilov, in terms of intervention at the federal level. President Shpektor was quite clear that he hoped that Kurilov, and the influence he brings to the Union through Norilsk, would help advance the federal Arctic law, "On the Arctic Zone of the Russian Federation."

> Oleg Kurilov is the first mayor to become vice-president of our organization. And the fact that Norilsk today leads the Union is no coincidence. I have been to Norilsk several times, and I see the positive changes in the city, and I attribute this dynamic to the competent actions of the local authorities. … Norilsk does a lot for Russia, so the congress will put forward a proposal to appeal directly to the President of the country to promote the law. (Strelnikova 2013)

The Union used its Norilsk Congress as an opportunity to reiterate its longstanding concerns regarding the federal plans for the Arctic region. The Union remarked in particular on the perceived shortcomings of the proposed Arctic Zone legislation, "On the Arctic Zone of the Russian Federation." The Congress was a timely opportunity, as it was synchronized with the federal approval process for the bill. The organization had already provided comments on the Arctic Zone legislation several times during its approval process, much of which was made public at the Congress itself. One of the Union's primary aims is to expand input from Northern localities in all Arctic legislation and decision-making.

"On the Arctic Zone of the Russian Federation" was intended to establish the legal mechanisms to implement legislation adopted in 2008, 2013, and 2014, which set state policies and objectives for the region. It draws particularly on earlier federal Arctic development and security policy adopted in 2013, the "Development Strategy of the Russian Arctic and the Maintenance of National Security for the Period up to 2020" (hereafter, "Strategy 2013") (Russia 2013b).[9] Strategy 2013 updates a 2008 policy, "Fundamentals of State Policy of the Russian Federation in the Arctic up to 2020 and Beyond" (hereafter, "Fundamentals 2008") (Russia 2008).[10] The land area defined in the Arctic Zone Executive Order also applies to a new state program adopted in April 2014, "The Socioeconomic Development of the Arctic Zone of the Russian Federation for the Period up to 2020."

Strategy 2013 is authoritative in defining federal objectives, interests, and activities in the region. The Arctic Zone Order calls Strategy 2013 the "main document of the strategic planning of socioeconomic development in the Arctic region for [the] fixed period of the program" (Russia 2013a). The document puts into action a longstanding federal goal of making the Arctic region its own entity to be federal managed as a "separate subject of state governance … [with] … a separate monitoring system" (Heininen, Sergunin, and Yarovoy 2013).

Together, these documents establish Arctic natural resource and economic development as priorities for Moscow. It is "a primary Russian goal through the creation of a favorable operational regime to be applied within its boundaries. In the period 2016–2020 the Russian Federation will transform her Arctic Zone into a leading strategic resource base in Russia" (Ostreng 2010). Strategy 2013 states that the government will actively support businesses operating in the Zone, primarily in the development of hydrocarbon resources (Strategy 2013). According to the Kremlin, expanding resource development in the Zone could do much to meet Russia's needs for oil and gas and other types of strategic raw materials (Strategy 2013).

In order to implement this development plan, the new legislation diversifies, at least on paper, regional decision-makers beyond the Kremlin, due to a lack of specialized knowledge of local economic development needs and capacity at the federal level. Strategy 2013 proposes an active role for local governments in regional governance, as well as public–private partnerships and foreign investment. The Kremlin "envisages an important role for regional and local governments as well as for private businesses" (Strategy 2013). Scholars note that this is at least partly because Russia lacks the requisite technology and resources to profitably exploit its own natural resources. After long hostility to foreign influence in Arctic business, the Kremlin is becoming increasingly open to foreign investment (Heininen, Sergunin, and Yarovoy 2013), although this trend stalled with the Ukrainian crisis and Western sanctions.

The Union's role in the governance of the new Arctic Zone is yet unclear, although, given the Kremlin's development goals, Moscow's apparent opening to local input likely has in mind important economic centers in the region, such as Murmansk and Norilsk, rather than smaller, less economically important Northern outposts or civic associations like the Union of Arctic Cities. Nevertheless, the Union appears to be taking full advantage of the opportunity to provide more subnational input in Arctic regional planning. It is leveraging the language in these federal Arctic policy documents that expands the role of subnational entities in regional decision-making.

At the Norilsk Congress, after pronouncing the Arctic Zone bill to be "long overdue" for approval in Moscow, Union delegates raised multiple issues with the draft federal law. They argued that several issues of particular importance to them are not adequately addressed by federal plans or funding, including endemic challenges such as housing shortages, poor quality housing, and the lack of professionals in the region ("S'ezd soiuza gorodov Zapoliar'ia i Krainego Severa"). Union delegates described the lack of professionally specialized in-migrants to the region as an "acute problem," noting that Northern localities lack teachers and medical professionals. Such highly trained specialists are difficult to recruit to inclement and remote Northern areas.

Kurilov, for example, voiced concerns that the draft law had not mentioned any rules regarding housing in the Arctic Zone. He gave statistics as to the gravity of the situation in Norilsk, although housing is a common problem not only in the Russian Arctic, but also throughout the circumpolar region. As a remedy, he asked Union delegates to consider supplementing the federal legislation with rules that would guarantee co-financing for the construction of rental housing in the Northern territories (Strelnikova 2013).

> Today, there are 865 residential homes in Norilsk, 10 of which have been deemed uninhabitable, and their residents have been relocated, while four of them are still inhabited. Every year, about 30 apartments are found to be uninhabitable. Meanwhile, the construction of new housing under the current conditions in Norilsk remains economically unprofitable, which hinders the development of the territory. (Strelnikova 2013)

Union delegates also argued that more local knowledge should be integrated in regional decision-making, critiquing top-down administrative decisions that failed to understand local economic needs. They argue that federal authorities are more concerned with military defense and resource exploitation. "[T]he draft of the proposed law does not meet the needs of the North and does not take into account its specificity" ("S'ezd soiuza gorodov Zapoliar'ia i Krainego Severa").

In particular, Union delegates argued that the expertise of local administrations should be better integrated into regional decision-making. They questioned the rigidly siloed decision-making, the "separation of powers between levels of government, and the problems of ensuring the local authorities' expenditure responsibilities" ("S'ezd soiuza gorodov Zapoliar'ia i Krainego Severa"). The Union also proposed some solutions to immediate issues, for example, that federal Arctic legislation should include co-financing from the federal budget for rental housing in the Northern territories. They also proposed the creation of a state corporation that will manage housing in the Arctic Zone ("S'ezd soiuza gorodov Zapoliar'ia i Krainego Severa"). Beyond these recommendations, the Union delegates made an official decision to appeal to Russia's Federal Assembly, the upper house of parliament, to propose changes to several federal laws that define the scope and responsibility of municipalities.

Many of these comments and suggestions had arisen during the Arctic Zone legislation working group meetings in 2013, where the Union of Cities in the Arctic and High North joined several other actors to make proposals, including the governor of Yamal, Dmitry Kobylkin, representatives of the Murmansk region, and the Russian Geographical Society. By March 2013 the bill was revised based on these proposals and submitted to the Ministry of Regional Development. Igor Shpektor, president of the Union, was a member of the bill's working group. He described a substantial Union role in the development of the Arctic Zone bill and in regional decision-making in general.

> We have worked on this law for 15 years. During this time we tried to convince each president of the necessity of its adoption. Yeltsin, Putin, and Medvedev have all worked on it, and now, finally, we are now on the home stretch. For northerners this law is extremely important. There a number of observations that need to be taken into account by the ministry of regional development. ... Practically, it will certainly help the people living in the Arctic zone of ... [Russia]. (Polyanskaya 2013)

The Arctic Zone Executive Order: Norilsk and Murmansk

Alongside this Union advocacy, the Union member cities of Norilsk and Murmansk lobbied on their own behalf for inclusion in the Arctic Zone, using Union membership as one important strategy. The establishment of the Zone through the 2014 Executive Order was originally intended to be part of the yet unratified "On the Arctic Zone of the Russian Federation." The Executive Order determines the geographic area where these policies will apply by "isolat[ing] and resolv[ing] the legal status of the Arctic zone" (Norilsk 2013).

The Federal Ministry of Regional Development, which drafted the broader Arctic Zone legislation noted, "[t]he purpose of the bill is to consolidate on a federal law the status of the Arctic zone of the Russian Federation, its borders and composition, and the legal mechanisms to implement the main objectives and directions of the state policy in the Arctic zone of the Russian Federation" (MRD 2013). The Federal Ministry of Regional Development was dissolved in late 2014, and its responsibilities spread out over several other ministries. The Federal Ministry of Economic Development has taken charge of much of the Arctic legislation.

The 2014 Executive Order defines an Arctic that spans several existing administrative regions: four federal subjects, three portions of subjects, and some separate municipalities. It includes the entirety of Chukotka, the Murmansk region, and the Nenets and Yamal-Nenets autonomous districts. It also includes portions of Sakha-Yakutiya, Krasnoyarsk krai, and Arkhangelsk oblast. This is very close to the areas proposed in draft legislation.[11] The zone is modeled on the definition set (but never adopted) by the Soviet Council of Ministers' State Commission on Arctic Affairs in 1989 (Laruelle 2014, 29). In fact, Russia has never defined a single Arctic administrative unit. Several different working definitions overlap today, defined alternately by climate, economic features, and other criteria. Although it is commonly understood that the political and geographic Arctic region is the area above the Arctic Circle, (lat 66.33° N), the area to be included in the Zone is complicated as some administrative regions cross this boundary and would be split into Arctic and non-Arctic zones.

One of the Kremlin's implicit goals in this Executive Order is to limit the extent of the formally defined Arctic. There are two reasons for this: first, to limit federal financial obligations, since territory within the new designation will have to be funded by special programs, and second, to break up potential political challenges by including only portions of pre-existing administrative regions.

Murmansk and Norilsk both pursued inclusion in the zone in order to capture regional funding and participate in regional decision-making, as federal financial obligations in the newly established Arctic Zone will be significant. According to the Regional Ministry, total funding for the program from all sources by 2020 would be almost two trillion rubles, which accounts for a third of the federal budget – but the economic slowdown of 2014–2015 made these financial commitments unrealistic. The Ministry also plans to attract more than one trillion rubles of private investment, including from large Russian companies ("Proekt gosprogrammy").

The Arctic Zone is one of several Russian regions competing for investment from the Kremlin; another is the war-torn North Caucasus federal district. The North Caucasus is slated to receive an estimated 235 billion rubles in federal funding by 2020 (Russia 2012) – so about 100 times more than for the Arctic region. In interviews, local officials in Norilsk are very conscious that the new Arctic region will be competing with other regions for funding, although an understanding of the implications of inclusion in the Arctic Zone appears to be limited.

In interviews, local leaders in one of Norilsk's satellite towns appeared to be unaware of the political or economic advantages of claiming Arctic status (Laruelle 2013).

Norilsk, Murmansk, and the Arctic Zone

The Kremlin hopes to disrupt potential political challenges by delimiting an Arctic region that cuts across pre-existing administrative districts. Federal funding and new investment in the region could potentially strengthen the political position of regional governors and city mayors at the expense of central authorities.

Splitting existing political administrative regions benefits Moscow, but creates several headaches for the regions that straddle the new border. The federal funding that accompanies the designation will mean some parts of regions will receive additional funding, and others not. Dividing regions also conflicts with long-standing conceptions of Arctic territorial extent. Local authorities from several Northern regions criticized drafts of the bill, noting that the "bill suggested many northern areas that have traditionally been regarded as 'Arctic' might be excluded from the definition." As noted in a *Barents Observer* article on August 5, 2013, territorial definitions of the Arctic in early drafts of the legislation caused some areas traditionally considered Arctic to be omitted from the new zone, including Arkhangelsk oblast and the Republic of Karelia. Several Union members who were excluded wholly or in part have lobbied for inclusion in the Zone, in order to benefit from the federal spending and potential political benefits attached to the new region.

The city of Norilsk and Murmansk oblast were particularly active in efforts to amend aspects of the new legislation. They successfully lobbied for their own inclusion in the proposed zone. This was Norilsk's second such attempt, as it was not included in the original Arctic Zone law that was submitted to the State Duma in 1998 (Strelnikova 2013). However, Regnum news agency reported on August 12, 2013, that the question of Norilsk's inclusion remained unresolved when this early version of the bill was submitted but not approved by the Duma. Norilsk and Murmansk intervened in 2013, as the proposed bill began the approval process anew. Murmansk oblast lobbied for the inclusion of the entire oblast in the Zone, rather than only selected areas and settlements.

These are two of the most economically and politically powerful Union members. Norilsk is the administrative heart of the Norilsk industrial region, the site of one of Russia's richest and most extensive mineral deposits.[12] Located on the Taimyr Peninsula in Krasnoyarsk krai, the region lies a few hundred miles east of the northward-flowing Yenisei River, which empties into the Arctic Ocean. The Norilsk industrial region (NIR) hosts the largest production center of one of Russia's premier, private natural resource companies, Norilsk Nickel Mining and Metallurgy Company. Today, Norilsk Nickel controls one-fifth of the world's nickel deposits, 20 percent of the cobalt, 3 percent of all the copper, and almost 45 percent of the world's valuable platinum group metals (Foek 2008).

Norilsk leveraged its political connections for its advocacy effort in 1998. At this time, neither Norilsk nor its port town of Dudinka was among the cities to be included in the Arctic Zone when the original legislation was submitted to the Council of the Federation. The city lobbied for its inclusion, asking Senator Andrei Klishas of Russia's Federation Council to advocate on its behalf. Klishas represents Norilsk as part of Krasnoyarsk krai. He has close ties with the industry as well, for he quit his position as president of Norilsk Nickel to run for political office in 2011.

Klishas was a savvy choice for an advocate. He is not only politically connected, but supports the type of public and private partnership that federal Arctic legislation is attempting to promote. He is also a longstanding advocate for Norilsk's role in regional development. For example, Klishas spoke to the media just before he left Norilsk Nickel in 2011, promoting the company's recently released development strategy through 2025. He described this plan as designed to work in concert with federal investment in the region to improve quality of life in Norilsk. The strategy calls for a larger proportion of its Norilsk infrastructural investment to go to local non-industrial investment in Norilsk, in order to improve living conditions in the city. He noted that Norilsk Nickel could indeed provide investment for Northern development, given the country's difficult economic position. He noted, however, that the company still expects federal regional support for business and non-business related investment. "I think that the federal leadership has come to the realization that the resources on which our country largely lives and develops are finite. And that in order to allow the growth of the economy in the future, we must be active in the North."[13]

In 2013, when the Arctic bill was revived, Norilsk again approached Klishas. The senator is a member of the United Russia party and spoke before the Duma in favor of Norilsk's inclusion in the Arctic Zone. This time, Norilsk's advocacy paid off. The city and its port of Dudinka were included in the proposed Arctic Zone resubmitted to the Duma on May 31, 2013. In his statement to the Duma Klishas said, "More than a third of the Krasnoyarsk krai, which I represent in the Federation Council, is in the arctic and sub-arctic climates. Accordingly, there is every reason to include certain areas of this subject in the Arctic Zone" ("Chast' Krasnoiarskogo" 2013).

In 2013, the governor of the Murmansk oblast also attempted to amend the Arctic Zone to include the entire Murmansk region. The original bill only included particular settlements above the Arctic Circle and defined as "Arctic" only land areas with direct access to the Arctic Coast. Murmansk oblast is located almost completely above the Arctic Circle, and its regional governor, Marina Kovtun, argued that the bill should include any federal subjects as well as municipalities that are located north of the Arctic Circle (lat 66.33° N) (Staalesen 2013). Kovtun addressed then-Prime Minister Dmitry Medvedev in a letter, noting, "We find it imperative to use as main criteria the federal subjects' and municipalities' localization north of the Arctic Circle" (Staalesen 2013).

Kovtun's argument for the inclusion of all municipalities of the Murmansk region in the Arctic Zone appealed directly to the Kremlin's declared view of the North as an important economic engine for regional as well as national development. She argued, in short, that all the territory within the Murmansk region should be included because all contain enterprises whose activities are directly related to the solution of development problems of the Arctic Zone, including Arctic Ocean ports (Murmansk oblast 2014a).[14] Russian analysts speculate that the inclusion of the entire region would strengthen Kovtun, as the governor of an already-powerful region (Staalesen 2013). This is likely a key reason why the oblast was split up in the original legislation, which was drafted by the federal Ministry of Regional Development.

Kovtun has continued her advocacy for a strong Northern voice in the management of regional affairs, even after the Executive Order was issued and Murmansk oblast fully integrated in the Arctic Zone. In November 2014, she met with Arthur Chilingarov, known in the West as the polar explorer who planted the Russian flag on the Outer Continental Shelf of the Arctic Ocean in 2007 to assert Russian primacy in Arctic sovereignty. Chilingarov is the Special Representative of the President of the Russian Federation for International Cooperation in the Arctic and Antarctic, the President of the Polar Association. Kovtun spoke with him at Murmansk International Business Week III, where he was the guest of honor (Murmansk oblast 2014b).[15]

Kovtun noted the importance of regional expertise in setting regional development strategy. She further stated that Arctic citizens and cities must work together to advance regional development themselves, since the future of the draft federal law on development strategy for the Arctic Zone was uncertain:

> We are not talking about the private interests of the Murmansk region. We are looking at a broader scope. We are talking about the interests of the country. We need a platform for dialogue, where each area could present their competitive advantage, because they each have their own advantages. Today each of the Arctic regions is fulfilling its special role in the development of this territory.[16]

According to a Murmansk oblast press release, Chilingarov agreed with Kovtun's message of Arctic solidarity in support of these goals, particularly given the fact that momentum for a regional development strategy at the federal level seems to have waned. He supported her "proposal on the need for all the Arctic regions to rally for the practical enactment of the Strategy for the development of the Russian Arctic, the implementation of which weakened after the abolition of the Ministry of Regional Development, which had been responsible for implementing the strategy."[17]

There has been some solidarity among Arctic Union members on the issue of inclusion in the Arctic Zone. Igor Orlov, the leader of Arkhangelsk, sent a letter to Kovtun supporting Murmansk's effort to include the entire Murmansk region in the proposed zone.[18] Arkhangelsk oblast's support was perhaps also self-serving – it, too, was afraid of losing its Arctic status, as the original legislation proposed its division. Indeed, only certain municipalities and smaller areas of Arkhangelsk oblast were included in Zone in the final Executive Order.

Conclusion: affirming power structures in the Arctic

The effectiveness of the Union's lobbying efforts is yet unclear, since the draft federal bill, "On the Arctic Zone of the Russian Federation," has not yet been ratified. This study suggests, however, that despite language in new Arctic legislation that provides a larger role for capital and subnational actors in regional decision-making, in practice, the Kremlin has been less and less responsive to input on Arctic governance from the Union. An example of this is its new, less responsive administrative Committee in the Federation Council for Arctic Affairs. However, this trend may change in the future. We may see more space for the participation of regional actors in Arctic policy if the Union continues to take advantage of new Arctic policy language. Ultimately, although the Union may have little sway with the Kremlin in its own right, its strongest role may be indirect, as a useful and effective platform for powerful members to lobby the Kremlin.

Economically and politically well-connected Union member cities appear to have had more success lobbying for changes in Arctic Zone legislation and to have Moscow's ear to some extent. The Executive Order designating the territory of the Arctic Zone has emerged, and it is clear that Norilsk and Murmansk were effective in at least ensuring their inclusion in the Arctic Zone, if not in their other points of advocacy. Both Norilsk and Murmansk are wholly included in the Zone, probably based on their political strength, which stems from their large roles in the resource economy in the Arctic. The Union as an entity, on the other hand, does not appear to hold the same leverage.

Russia's Arctic project is a new manifestation of center–periphery power struggles that recalls Russia's long history of political and economic tensions between the central state and regions. The activities of the Union and its members suggest new relationships between the state and business that challenge the unilateral federal role in setting the Arctic development agenda. They shed light on how longstanding center–periphery political tensions in Russia extend into the Arctic.

Most Arctic legislation is directed by central authorities and reflects federal interests. Indeed, the Arctic Zone legislation and the establishment of the Union of Cities were initiated "from above," as the Kremlin attempts to advance federal interests in the Arctic and disrupt potential challenges to centralized power. Further advocacy by the Union and its members was conducted *through* central authorities, indicating the overall success of the Kremlin's ongoing efforts to strengthen its power over Arctic decision-making. However, the success of some economically strong cities in challenging the composition of the region suggest that urban centers that the Kremlin considers important either for their resources or strategic value can have a role in shaping the Kremlin's plans for the region.

Indeed, the Kremlin faces the problem of local power accretion. The problem has grown as regional boundaries acquired bureaucracies and strong local executives. The Union is attempting to regain local control. Although the Union did not directly challenge the strong central role of the Kremlin in developing and securing the Arctic, it had some success in changing aspects of regional policy.

It lobbied for changes in regional policy as a collective actor with potentially significant pull. Its members include cities with large amounts of natural resources with significant economic and political strength.

Although the Kremlin's efforts are intended to tame the regions, the Arctic Zone legislation and the Union itself undermine unilateral decision-making in important ways. The Union's advocacy and assertions of local expertise in regional decision-making suggests that this entity and its members have become what Goode describes as "vehicles for ... [a] ... developing sense of regional distinctiveness and, eventually, distinctive regional identities" (Goode 2011, 56). Indeed, organizations like the Union are vehicles shaping the characteristics of the Arctic region. Their intervention creates a region with distinctive forms of governance reflecting a strong regional identity.

Notes

1 Ukaz Prezidenta RF of May 2, 2014, No. 296, "O Sukhoputnykh territoriakh arkticheskoi zony Rossiiskoi Federatsii," http://graph.document.kremlin.ru/page .aspx?3631997.
2 "Ob Arkticheskoi zone Rossiiskoi Federatsii."
3 Press Center note May 14, 2015. RAO/CIS, www.rao-offshore.com/index.php?option =com_content&view=article&id=679%3A2015-05-14-13-06-33&catid=5%3 Anews&Itemid=54.
4 Soiuz gorodov Zapoliari'a i Krainego Severa.
5 The list of Union cities and regions is provided on the website: http://krayniy-sever .ru/?page_id=21.
6 http://krayniy-sever.ru/.
7 See the Union Charter at http://krayniy-sever.ru/?page_id=9.
8 http://krayniy-sever.ru/?page_id=19.
9 "Strategiia razvitiia Arkticheskoi zony do 2020 goda."
10 Fundamentals 2008 was approved by President Dmitry Medvedev, on September 18, 2008, No. Pr-1969; Strategy 2013 was approved by Putin on February 8, 2013, No. Pr-232.
11 The Regional Development Ministry proposed the following demarcation in draft legislation: "It is proposed that the following regions – the Murmansk region, the Nenets Autonomous Area, the Yamal-Nenets Autonomous Area, the Chukotka Autonomous Area, the Sakha (Yakutiya) Republic, the Krasnoyarsk krai and the Arkhangelsk region – be included in the Russian Federation's Arctic zone." Furthermore, it is also proposed that Chukotka, the Murmansk region, the Nenets and Yamal-Nenets Autonomous Areas be completely included in the national Arctic zone. The ministry statement continued, "Five of the 35 municipal entities in Yakutiya, three of 61 municipal entities in the Krasnoyarsk Territory, as well as seven of the Arkhangelsk region's 26 municipal entities will be included in Russia's Arctic zone as well." ("Proekt gosprogrammy" 2013).
12 This research focuses on the city of Norilsk, as the largest city and governing authority in the Norilsk industrial region. I refer to Norilsk in this research to represent the NIR as a whole. The Norilsk municipality (with Norilsk Nickel's influence) makes most of the administrative decisions that affect the populations in the region's smaller satellite towns of Talnakh, Kayerkan. and Snezhnogorsk.
13 "Andrei Klishas: Noril'sk – mesto, gde mozhno ne tol'ko rabotat', no i khorosho zhit'." Accessed November 20, 2015, http://nia14.ru/andrei-klishas-norilsk-mesto-gde-mozhno-ne-tolko-rabotat-no-i-khorosho-zhit.

14 http://new.gov-murman.ru/info/news/884/?sphrase_id=71741.
15 "Gubernator Murmanskoi oblasti Marina Kovtun prizyvaet vse arkticheskie regiony k konsolidatsii." November 18, 2014. http://new.gov-murman.ru/info/news/48539/?sphrase_id=71739.
16 Ibid.
17 Ibid.
18 "Arkhangel'skaia oblast' podderzhala Murmanskuiu v voprose ob Arkticheskoi zone RF." *Arctic Info*, August 1, 2013, www.arctic-info.ru/News/Page/arhangel_skaa-oblast_-podderjala-myrmanskyu-v-voprose-ob-arkticeskoi-zone-rf.

3 Reluctant entrepreneurs of the Russian Far North

Aimar Ventsel

Research into private businesses in post-Soviet Russia started with the collapse of the Soviet Union. While more than two decades have passed since the USSR disintegrated, many scholars still regard private entrepreneurs as participants in an economic situation that is new to them. This chapter focuses on a specific segment of private entrepreneurs who I call "reluctant entrepreneurs." These people entered into business reluctantly in the early 1990s, when salaries in state jobs were not paid. These entrepreneurs also plan to return to a public-sector job at a later time in order to collect their social benefits. Therefore, although most of them have been engaged in private business for some time, they see it as a temporary activity. In their activities, entrepreneurs are led by local social norms and the policy of risk aversion. This attitude takes place within an economic climate where the state largely ignores informal and gray-market economic activities, and where enterprises are kept small, non-innovative, and diversified.

The defining line between the socialist society and "what came next" is apparently when the former "economy of shortage" was transformed into a market economy, where goods became widely available, and entrepreneurship was allowed to exist legally (see Verdery 1993, 1996). The collapse of the socialist planned economy in Eastern Europe was hailed not only as an economic transformation in Western thinking, but also as the emergence of an entrepreneurial class, associated in the Weberian tradition with the appearance of a fundamental need for innovation, free thinking, and democracy (see Davidova and Thomson 2003; Estrin, Aidis, and Mickiewicz 2007; Gibson 2001; Peng 2001; Wegren 1998, 2000). Since then, entrepreneurship in Russia has taken on different forms and is more often related to the shadow economy, struggle, and crime than with any kind of free thinking or democratic ideology (see Cartwright 2001; Humphrey 2002; Kuznetsov and Kuznetsova 2005; Seabright 2000; Tarasov, Egorov, and Kulakovskii 2013; Ventsel 2005). The private economy, however, continues to thrive and plays an important role in peoples' everyday life in Russia.

Reluctant entrepreneurs often took up their new occupation not because they dreamed of becoming entrepreneurs, but because *biznes* seemed to be their best option for survival. With a few exceptions, these people did not plan to be engaged in entrepreneurship for their whole lives, but wish to return to the public sector or to divide their time between their enterprises and a state-paid job. Moreover, their views on profit and expansion might contradict Western theories on leading a business successfully. As I show, such entrepreneurs are not an anomaly, and their strategies and logic are fully understandable given the current state of the market economy in Sakha (Yakutiya).

The Republic of Sakha (Yakutiya) and its economic environment

The Republic of Sakha (Yakutiya) is the largest territorial unit of the Russian Federation and belongs to the easternmost administrative unit, the Far East federal district, which contains nine different territorial units (Map 3.1). The republic is a large but sparsely populated territory covering more than three million square kilometers, where slightly less than a million people live, 55 percent of which are the titular ethnic group, the Sakha. The republic is famous for its diamond resources, producing 30 percent of the world's diamonds and almost 100 percent of Russia's diamonds. In addition to diamond mining, the Republic of Sakha has significant resources including gold, gas, oil, precious metals, coal, and timber. Natural-resource extraction is the domain of big companies, as is large-scale construction as well as air and water transport. Some 73 percent of the population in the Republic of Sakha is urban.

It is not entirely clear how officials define medium or small enterprises, but in 2009 in the whole of the Russian Far North there existed 181,514 small enterprises (Stepanova and Nogovitsyn 2011, 49). Approximately 80 percent of these were "microenterprises" or enterprises that hired up to 15 employees on a permanent basis. It must be mentioned that with regard to the number of enterprises per capita, the Far North is slightly below the average Russian level. In terms of regional distribution, about 47 percent of small enterprises are located in Primorskiy krai, 19 percent in Khabarovsk krai, and only approximately 8 percent in the Republic of Sakha (Yakutiya) (49–59). Small enterprises tend to have very low investment and profit levels, and 60 percent of Russian Far North small enterprises are engaged in whole and retail sales (Egorov 2006, 192; Stepanova and Nogovitsyn 2011, 65).

In 2010, there were 115 medium-size enterprises and 4,952 small enterprises registered in the Republic of Sakha; from these 4,029 were classified as microenterprises or enterprises that employ up to 15 employees. In the year 2009, small enterprises contributed only 6.6 percent of the regional GDP but gave work to 40 percent of the local workforce. The particularity of the region is that 80 percent of these enterprises are concentrated in four cities: the capital, Yakutsk; Neriungri, famous for its coal industry; Mirnyi, the diamond industry's center; and Lensk. From the regional enterprises only 400 are agricultural. The statistics show that in Sakha 44.4 percent of small enterprises are engaged in wholesale merchandise, 22.7 percent construction, 6.8 percent real estate, and 5.7 percent transport (Figure 3.1). The role of small-scale enterprises in different spheres can be high: 94.3 percent of services (*bytovye uslugi*) and 100 percent of small bus transport belong to small enterprises as does the majority of tourism, clothing repair, cargo transport, car and truck repair, and other services.

Map 3.1 The Republic of Sakha-Yakutiya.

Source: Anderson, David G. 2000. *Identity and ecology in Arctic Siberia: the number one reindeer brigade*. Oxford: Oxford University Press. p. 8.

The low profit levels of such enterprises can be explained by the underdeveloped infrastructure and high energy costs in the region. Local economists have highlighted one peculiarity of small-scale enterprises: their need for highly skilled specialist labor is much lower than their demand for unqualified manual labor. Notwithstanding several federal and regional programs, the state has been unsuccessful in developing "innovative" and modern high-tech small enterprises. For example, when in 2009 the state start-up enterprise support foundation *Start* announced grants for innovative new firms, 33 enterprises from the Republic of Sakha applied, but only nine of them qualified as innovative and received grants. In general, the statistics show that between 2004 and 2009 only 29 innovative small enterprises were established in the republic (Stepanova and Nogovitsyn 2011, 56–57). As some local scholars argue, big companies with better financing and market prospects dominate these spheres already (59).

This chapter draws on the fieldwork I conducted in the Republic of Sakha in July and December 2013 and in July 2014. Each time, I spent three to four weeks in the region, conducting research mainly in the capital of Yakutsk, but also in villages near Yakutsk and in the Verkhoyansk district (approximately 1,000 kilometers northeast from Yakutsk). During the fieldwork I recorded 10 interviews with different entrepreneurs whose enterprises were classified as small or microenterprises.

Figure 3.1 Building site of private garages. Renting garages for car owners during the winter is a substantial source of income for private entrepreneurs.

Additionally, I talked to various other entrepreneurs, visited their enterprises, and observed how they operated. All but one entrepreneur had started their business out of necessity in the 1990s. One woman was the spouse of a wealthy business-man whose husband funded her bridal shop "because I always dreamed of such a store." Nevertheless, she also perfectly fitted the ideal, sharing similar attitudes and behavior patterns with other subjects. During the time of the fieldwork, my respondents were engaged in various sectors: fishing and hunting enterprises, food shops, laundries, construction firms, tourist ventures, and a hair salon. Some of my informants had different small enterprises, none of them had studied the profes-sion they were engaged in at the time of our encounter, and for all the respondents their current enterprise was not their first but rather a third or fourth enterprise. All but one were ethnic Sakha (one was Armenian), half of my respondents originated from a village, roughly 70 percent were women, and all of them were in their 40s.

The beginning

Existing studies on private entrepreneurship in Eastern Europe rarely disclose what motivated the actors to start their *biznes* in the first place. The literature often mentions "when" people became involved in business, but that fact is sel-dom accompanied by the question "Why?" Nevertheless, the general impression of the academic literature is that individuals took up the occupation more or less voluntarily and even with some degree of enthusiasm. In general, one expects that a person who enters into a small business possesses the wish to be engaged in entrepreneurship and has some vision or strategy as to how he or she will develop the business. Indeed, studies of entrepreneurship in post-Soviet Russia and grounded in anthropology (Hann 2000; Köllner 2013), economics (Berkowitz and DeJong 2001; Estrin, Aidis, and Mickiewicz 2007; McCarthy, Puffer, and Shekshnia 1993), and sociology (Barsukova and Radaev 2012; Bogatyr' 2013; Sakaeva 2012) indicate indirectly that the actors saw the new market economy as an opportunity. Another popular argument in academic literature is the "path dependency" argument, an assumption that entrepreneurs have role models, have been active in the business previously, or that new post-socialist enterprises follow the footsteps of similar state companies of the socialist period (Alanen et al. 2001; Nee 1991; Pickles and Smith 1998; Pissarides, Singer, and Svejnar 2000; Stark 1992).

The group studied in Sakha had similar motivations to switch their occupation. Most informants related a version of the same story: "When in the 1990s wages were not paid I had nothing else to do but to open my own kiosk. We sold everything, day and night!" All but two informants entered into entrepreneurship via the kiosk economy or selling various goods in the markets, embracing the fast spread of *biznes* in Russia in the 1990s and the beginning of the 2000s. The majority of these people had worked, until that period, in low-paid, public-sector jobs, which were the first to go unpaid or received extremely poor salaries in the new economic conditions. Among my respondents were former teachers, library workers, day care providers, university teachers, accountants, and mid-level clerks in various state enterprises. One respondent explained,

> When I graduated with a history degree in 1996 I went to search for work. In the school they offered me a salary of $10 or something like that. I thought I am not going to work for that money. So, I was eagerly looking for some normal pay with my fresh university diploma. At the time my brother started to trade at the market and asked me to be one of his sellers. So all this started and by now I have not worked a single day as a historian.

Others had quite a substantial career behind them. As one woman told me: "Our enterprise was broke. We worked months without any salaries. We had a problem with how to feed our children. My husband worked in several jobs but that was not enough. So I decided to open the kiosk."

There is one divergent story in my data. One of the informants was forced to take over a fishing and hunting enterprise because her mother had died and left the enterprise behind. "My mother died and I had no choice. Somebody had to take control. That was an enterprise she had built up and we could not abandon it!" In answer to my question as to why she felt obliged to take over the leadership, she replied: "My mother had established the enterprise, how could I give it up? Moreover, all the people working for the company are our relatives in the North [in the village of Tiksi on the coast of the Arctic Ocean]. How will they receive their salaries if the company ceases to exist?"

Morality of the business

The last quotation introduces another phenomenon related to the reluctant entrepreneurs in the Russian Far North: the moral dimension of the business that follows these entrepreneurs during their active career. Different moralities seem to play a significant role in shaping the business ideology. An Orthodox businessman in Tobias Köllner's (2013) book struggles to observe Christian principles in his relations with his employees, the state, and customers; but Far North small-scale entrepreneurs follow norms that are more relatable. These norms are embedded in local kinship ideology, a notion of social value and morality. Different theoretical approaches to morality agree that the perception of what is moral or not is anchored in cultural values (Brandtstädter 2003; Estrin, Aidis, and Mickiewicz 2007), although according to some scholars morality should be publicly demonstrated through behavior and the decisions made (Robbins 2007).

As mentioned, entry into the world of entrepreneurship for my informants in most cases was involuntary, the incentive being a need to seek alternative income, not the desire for self-realization. This entry did not exclude, however, following certain moral principles that remained consistent. It was not only providing income for relatives and demonstrating kinship solidarity that was important for Far Northern entrepreneurs in the early 1990s.

> Working in a kiosk is quite dull. You just sit in there and sell what people want. I never hired young men (*parnei*) because this is not a man's work to sit and sell cigarettes. Girls are more able to adapt [to the situation], this work is more suited for them. Moreover, girls have a better sense of discipline. They do not drink heavily and then skip the next day. For a girl, it is a good opportunity to earn some money when they have just arrived to the city and need some finances. But this is not a job for a guy to start a career.

In this and similar statements, practical reasons merge with local perceptions about gender and masculinity. In the Russian Far North, where life is still tough for most working people, the man is foremost a breadwinner. On the one hand, according to informants, men are less reliable and controllable than women. Yet work in a kiosk contradicts the local perception of masculinity, where a man should earn money through heavy physical work and not sitting in a small booth.

These examples confirm that the concept of the "business culture" – that local social and cultural norms affect the way entrepreneurs shape their business – also has its relevance in Russian Far North (see also Ledeneva 1998, 2000, 2006; Nee 1991; Peng 2001; Puffer, McCarthy, and Boisot 2010). However, social norms and moral values affect not only the way in which the company is linked to formal and informal networks, as is the main focus of the analysts of the role of culture in business, but goes further.

Perceptions of gender are important for hiring employees, as is the obligation to provide income for relatives. Moreover, many respondents confessed that they prefer to hire pensioners. From the practical side, pensioners in Russia usually agree to work without a contract, are more disciplined, and accept lower pay. This practice was also, however, seen as a form of charity by some entrepreneurs, because old people were able to earn additional income to supplement their low state pensions.

The entrepreneurs were sometimes, nevertheless, making decisions that were clearly unprofitable, especially in the short term. Most of my respondents who had food shops stated that they refuse to sell alcohol. In many cases, they also had another store in a rural area or were supplying stores that were run by their close relatives (Figure 3.2). I was told, "We do not support it [alcohol trade]. We support a healthy way of life [ZOZh in Russian, a common abbreviation for *zdorovye obrazy zhizni*]." It is true that alcohol consumption in the rural areas of Sakha is a big problem. In recent years the government has implemented an aggressive (although ineffective) policy of limiting alcohol sales, banning alcohol use from public places, and promoting anti-alcohol propaganda. My fieldwork in Arctic villages in summer 2014 discovered that many shops, even entire villages in remote areas, have banned the sale of alcohol following a decision by the local community, not as the result of any government initiative. When an economic rationale comes into conflict with social norms, the social norms prevail. In the following section I will place the "business culture" into a broader framework.

Figure 3.2 Typical privately owned food store. Village of Batagai (photo by the author).

Explaining reluctance

As is apparent from various anthropological, sociological, and economic studies, the newly emerging entrepreneur class may not behave according to the rules of transparent trade. However they possess a certain amount of the required mindset that one expects from people who have devoted their life to building up and developing their business: a desire for expansion, a keen eye for innovation, and a rational, cost-effective strategy. This is, of course, true for small entrepreneurs in the Republic of Sakha, but only to a certain extent. In this section I will refer to certain strategies and general socioeconomic constraints that elaborate on this phenomenon.

One day I was visiting a friend, a successful manager and famous local showman. Suddenly his office door opened and a modestly dressed elderly woman unceremoniously marched in. She barely acknowledged me and started to talk to my friend in Sakha, something Sakha people usually avoid doing in the presence of foreigners, assuming that this is impolite because usually we do not speak Sakha. She was asking my friend to look for a man's hairdresser for her small hairdressing salon. My friend explained to me that the woman was from his village, where he has several *biznes*. The lady refused to give me an interview but did agree to speak with me off the record. As it turned out, she had moved to the city in the early 1990s, when salaries remained unpaid in the villages. She started with kiosks and soon opened a small enterprise, initially producing toilet paper. After some years she sold her business and invested the money in some food shops. The accumulation of income was not used for enlarging her initial business but for opening a laundry. When the laundry became successful, she opened a hair salon. In the discussion with her, and after getting additional information from other people, I understood that she managed her enterprises herself, using only the occasional help of her children.

A similar pattern became apparent with other entrepreneurs: they relied on informal kinship networks and were not very interested in expanding their enterprise, but rather the diversification of risks by investing into an unrelated sphere (see discussion of the "recombinant property" and "risk aversion" strategies of post-socialist enterprises in Stark 1996). The reluctance to grow and develop the otherwise well-functioning enterprise was also signaled by the near invisibility of the shops. Most enterprises I visited had either extremely modest signs on the street or nothing at all. As the owner of an unmarked laundry explained to me, her customers come to her by word of mouth. People who were engaged with supplying village people with furniture and domestic tools relied upon a network of customers and had no formal catalogue or website.

The mistrust of formal institutions and a preference for informal networks in the post-socialist economy is widely known. In general, small-scale entrepreneurship is always related to an informal economy, tax avoidance, and illegal wages (cf. Beneria 1989). Others argue that entrepreneurs rely on informal networks until the formal structures have developed into fully trustworthy institutions (cf. Kuznetsov and Kuznetsova 2005). The literature on post-socialist entrepreneurship often emphasizes that the informal networking and "blurred boundaries" exist due to the new economic situation after the collapse of the socialist planned economy, where businessmen entered a field previously unknown to socialist people (see also Pissarides, Singer, and Svejnar 2000; Torsello 2003). That assumption was certainly true in the 1990s, but more than 20 years after the collapse of the socialist economy this argument seems unconvincing. The small-scale private economy in the Russian Far North remains as far from the Western entrepreneurial ideology (cf. Kuznetsov and Kuznetsova 2005; Tkachev and Kolvereid 1999) as it did two decades ago. Many studies stress that the institutional environment explains the low level of entrepreneurial development in Russia (cf. Estrin, Aidis, and Mickiewicz 2007), and this is true, but in a different way.

Namely, in the Russian Far North, small enterprises have an unexpected freedom in their activities. When I asked one of my informants about her relationship with the state tax office, she told me: "They do not care! Here the tax office is after big companies. They get their income from fines and because the sums they can fine us are small, they are really not interested in fining us. Usually they do not care, even if someone informs them about some violation." This was confirmed by a professor of economics at a local university, a former minister of the economy of the republic: "The tax officials get rewarded according to the amount of money they return to the republic budget." Local scholars also criticize the regional government's emphasis on developing large industry and neglecting small and medium enterprises (Egorov 2006, 188). Still, the reality shows that local government structures have little interest in controlling this segment of the economy. In a situation where nearly 60 percent of the Yakutsk population has problems with finding suitable work, most small entrepreneurs are engaged with reselling imported goods and the spending capacity of the local population is low (192). Entrepreneurship does not involve a risky "guerrilla strategy," but rather enables people who are interested in satisfying the elementary needs of the population. Such a situation is not only motivated by "gray individuals" or people who had experience with the shadow economy during the Soviet era to start their own *biznes* (Peng 2001, 96), but also by former teachers, nurses, bookkeepers, or day care providers.

The reluctance of people to follow "normal business rules" was especially obvious when it became clear that entrepreneurs do not form a community, as one would expect. All the respondents told me that they have little interest in knowing their competitors and that they do not communicate with them at all. "I never communicate with other owners of wedding garments shops," one lady told me. "Most of the shops are in the western part of town, I am here in a center. They sell Italian clothes, I order my stuff from China. I spend my time with former colleagues, mostly teachers" (Figure 3.3). Another person exclaimed, "No, I do not know other owners of laundries. I know that there exist 28 laundries in the town and I have met the owners only once when two years ago we were called to the city government for a meeting. Otherwise, I have nothing to do with them!"

Figure 3.3 Wedding garments shops owned by one of the entrepreneurs cited in this text
(photo by the author).

Conclusion

Anthropology has the potential to "contribute to our understanding of economy" by finding rationalities behind seemingly irrational or illogical economic behavior (Gudeman 2008, 14). When economists interpret economic success and development in terms of transparency, verifiable statistics, and following the logic of the market economy (see Davidova and Thomson 2003) then, in reality, people's agency gives different meaning to their strategies and practices. In Hann and Hart's reading of Parry and Bloch (1989), and Durkheim (1960), money and markets must be "impersonal and asocial," but the relations between people or a personal relation to society are, on the contrary, personal (Hann and Hart 2011, 94).

Personal relationships in the world of business include a variety of social norms and strategies that affect "economically rational behavior." The entrepreneurs of the Russian Far North demonstrate that economic practices can be better understood when looking at the economic environment in the region and people's biographies. In the case of reluctant entrepreneurs, entry into business was often involuntary, a step to choose the best option among the bad possibilities available. Ethnographic interviews with people offered alternative explanations on the concepts of *blat* (influence), "blurred boundaries," and "informal structures in the economy."

While there is a widespread understanding that private business maintains affiliations with the state structures in order to profit from that connection, my research shows that this connection is also needed in order to withdraw from private entrepreneurship "when the time is over." All but one of my informants told me that being a "state employee" is significantly better than being a self-employed entrepreneur. In a state job, one receives a stable income, sick leave, and (very important in the Far North) annual paid travel inside the Russian Federation during the holidays. Notwithstanding the general rules of the Russian Federation, non-government enterprises hesitate to cover these costs by manipulating work contracts or refusing to have any contract at all. For self-employed company owners, the money would come from his or her own pocket. Therefore, the reluctant entrepreneurs maintain a connection with the public sector in order to be secured against an unexpected backlash in the private sector and to have easy entry into the public sector when retirement age is nearing. In fact, some older informants expressed their plans to close their business and seek state employment approximately 10 years before their pension age.

Considering the future plans, biographical facts, and business philosophy of my respondents, their reluctance to follow "normal" business rules is explained by the fact that they see their activity as temporary. Their entry into the business world was made possible by the framework of a local economy where the demand for innovative enterprises with highly skilled managers is nearly non-existent, and where state institutions show little interest in controlling small enterprises. Former teachers or nurses were able to use their organizational abilities and networks in order to establish themselves as entrepreneurs, but they avoid entering the spheres where substantial specialist knowledge is needed. In a similar way, they are reluctant to let their enterprises grow beyond the point where they need to hire educated managers. Instead, the "reluctant entrepreneurs" often prefer to exist in a gray area and diversify their activities to avoid economic risks.

Part II
New mobility patterns

4 Migration cycles, social capital, and networks

A new way to look at Arctic mobility[1]

Nadezhda Zamyatina and Aleksey Yashunsky

By its very nature, the Arctic is a mobile, unstable, and pulsating region. The boundaries of the polar ice, tundra, and forest limits are constantly changing; many animals are characterized by seasonal migration, and human settlements are often temporary. Life in the harsh conditions of the Arctic relies on a delicate balance that is highly sensitive to outside changes. Changes in the external conditions usually trigger internal movement: the boundaries of natural areas shift depending on the global climate change, many animals migrate depending on the season of the year, and the aborigines follow the animals.

While movement takes place due to the changes in external conditions, the particular trajectory is determined by internal circumstances, by the specific structure of the Northern space in which some directions of movement are more preferable than others. The animals have certain paths of migration; the borders of natural areas are displaced along the shapes of the relief: the forest penetrates northward along the river valleys and lowlands, while tundra penetrates to the south along the watersheds.

The Arctic environment offers many analogies and metaphors to describe its migration situation. Thus, in recent decades some areas of the Russian North have experienced the ebb and flow of the population, largely associated not as much with the internal situation in these areas, but with the more general socioeconomic processes taking place in the country as a whole. In the summer of 2014 we saw the clearest example of exogenous causes of migration in the Arctic, when large numbers of migrants from war-torn eastern Ukraine ended up in the north of Russia, thousands of kilometers away from the "hot spot," prior to the official relocation of refugees throughout the Russian Federation.[2] Just as the topography of the earth's surface structures the physiographic processes and defines the channels for movement and networks in the Arctic, social processes, particularly migration, are also subject to specific forces that structure their deployment in space. Migrants from the Donbass ended up in those Northern regions, which were relatively "closer" to Donbass Basin in this structured and anamorphic space than would be expected due to the physical distance.

Shifting from metaphors to explanations, we can apply the concept of *proximity* (Torre 2008, 2011; Torre and Gilly 1990; Torre and Wallet 2014; Boschma 2005). The concept of proximity is increasingly used in economic geography, regional economics, and sociology. The notion of "proximity" is close to the traditional understanding of distance in classical geography, but it is a wider notion that allows geography to work with an explanation of phenomena at the contemporary level, in the face of declining transportation costs and developing global networks. Proximity is similar to distance in the sense that both concepts describe the background, "environmental" conditions for the appearance and maintenance of interactions between two objects. In the process of working with distance, we argue that close objects are more likely to be related to each other – this is based on the classic theory of central places, among others. When we shift to proximity, we note that the interactions between objects are determined not only by physical distance, but also by their social, institutional, and other similarities or differences; we apply the concept of distance as a metaphor – proximity means the distance in the conventional social, institutional, and others kinds of spaces. Thus, we focus on a multidimensional space in which physical distance is only one of various possible measurements.

Usually, proximity is recalled in relation to reduced transport costs, communications networks, and the "death of distance" as a whole.[3] However, proximity plays a special role in the Arctic – precisely because of the great difficulties associated with overcoming physical distance (e.g., Laruelle 2014). Despite the spread of air transportation and all-terrain vehicles in sparsely populated and low exploration areas, the Arctic is a special, anisotropic space where the different sections and directions of movement are not equivalent to each other. In this sense, the space of the Arctic has a very complicated, crossed topography that determines the permeability of space: moving in some directions (e.g., along the river) is still much easier than in others. In addition to infrastructure, other forms of proximity – social, institutional, organizational – play a huge role in transforming the space of the Arctic, particularly regarding the interaction between actors in the Arctic and actors at distances far from it. Strictly speaking, if there was no connectivity within the framework of certain types of proximity, the Arctic would not be as connected with the outside world as we see it in reality.

Meanwhile, close connectivity with other regions of the country, the possibility of frequent movement between the Arctic and non-Arctic areas (in the Russian North, residents refer to non-Northern parts of the country as the "mainland" or "earth," or the "Great Land") significantly improves the quality of life in the Russian Arctic. When speaking about living conditions, the Northerners constantly mentioned the importance of communication with the "mainland." There are two main types of connections. One provides a temporary geographical proximity to other areas of the country. The second can be described as deferred proximity: a hope for future migration "to the mainland." This chapter explores how these forms of geographical proximity (temporary and deferred) are determined by social and other forms of organized (super-geographical) proximity – primarily social networks.[4]

Objectives and research material

Many migration researchers have already noted that out-migration from the North is configured by social networks or other types of proximity. For example, in Magadan oblast in the 1990s researchers observed "two trends in the formation of migration flows: the inhabitants of the oblast tried to go to areas from where they had come, and where there were relatives of theirs (Ukraine, Rostov oblast, and the krais of Krasnodar, Primorskyi, Stavropolsk, and Altai), or to the areas that provided construction of housing for the Northerners (Moscow, Vladimir, and Tula oblasts)" (Soboleva and Melnikov 1999).

Our current research seeks to provide a detailed assessment of how the migration of young people from the North is determined by social ties with the "historic homeland" of migrant families. We hypothesized that, when leaving from the North, young people flock to the places they, their parents, or their grandparents once left. In other words, we consider exodus from the North as a continuation of a family cycle, "migration to the North – migration from the North," which takes between one and three generations to complete. We hypothesize that such internal family ties form a spatial projection of this cycle.

Due to the absence of official statistics on "city-to-city" migration, we used two alternative sources to study the migration of young people from the Northern cities of the Krasnoyarsk krai. First, we used data on the migration intentions of senior schoolchildren surveyed in the three cities (242 persons in Norilsk, 66 in Dudinka, and 43 in Igarka,[5] which form 8.8 percent, 19.8 percent, and 43 percent of the total number of 10th and 11th grade students in the respective cities).

Second, we used personal data, such as school and university attended, place of residence, and age. This data is publicly available on the online social network "VKontakte" (Zamyatina 2012; Yashunsky and Zamyatina 2012; Chekmyshev and Yashunsky 2014). Using a special computer program, data was collected on all registered users of the network who went through educational training in the schools of the studied cities and who were in the age range 20–29 years at the time of the study. With the methodology developed for data collection through the online social network VKontakte, for the first time (in Russia) we were able to undertake an almost continuous analysis of migration of young people from the cities of Norilsk, Dudinka, and Igarka (and for comparison purposes from some other cities of the North: Noyabr'sk, Magadan, Gubkinsky, and Muravlenko). A total of 11,618 personal profiles were collected in Norilsk, 1,854 in Dudinka, and 789 in Igarka. Based on the estimates, the coverage of the age cohort of the population in the respective cities is at least 80 percent.

Migration in social proximity networks: how youth migration is configured by family history

A detailed study of the results of the survey confirms that the choice of some cities is almost exclusively defined by return migration: children born in the North go to the birthplace of their parents, in other words, a return "to the land of their ancestors" is performed through a generation. In Table 4.1 these are those cases where Norilsk dwellers of one or two generations plan to go to Kazan, Baku, Nizhny Novgorod, and Abakan. Most often, recent migrants expect to eventually return to provincial centers: they return to where they were born, to the nearest major city, or to their relatives in another city. For example, our survey sample includes a girl born in Nefteyugansk (Khanty-Mansiysk Autonomous Okrug) who plans to move to Ufa, where her parents used to live.

Table 4.1 Features of migration preferences of different groups of Norilsk and Dudinka dwellers, based on the degree of rootedness

Cities leading in migration	Norilsk (Dudinka dwellers) of second generation		Norilsk (Dudinka dwellers) of first generation		Recent migrant		Total sample (taking into account those whose residence period in Norilsk and Dudinka is unknown)	
	People	%	People	%	People	%	People	%
Total number of the group	70	100	151	100	62	100	307	100
Plan to remain in Norilsk	4	5.7	7	4.4	3	4.8	14	4.6
Are assured that will leave for another city	45	64.3	115	72.8	39	62.9	208	67.8
Preferred destination cities for migration	St. Petersburg	20.0	St. Petersburg	16.4	St. Petersburg	11.3	St. Petersburg	17.3
	Krasnoyarsk	15.7	Krasnoyarsk	14.5	Krasnoyarsk	9.7	Krasnoyarsk	14.0
	Moscow	7.1	Moscow	8.8	Baku	4.8	Moscow	8.5
	Novosibirsk	4.3	Baku	2.5	Moscow	4.8	Novosibirsk	2.9
	Astrakhan	2.9	Kazan	2.5	Astrakhan	3.2	Baku	2.3
	Ivanovo	2.9	Belgorod	1.9	Barnaul	3.2	Kazan	2.0
			Nizhny Novgorod	1.9	Ufa	3.2	Nizhny Novgorod	1.6
			Abakan	1.3			Belgorod	1.6

(continued)

Cities leading in migration	Norilsk (Dudinka dwellers) of second generation		Norilsk (Dudinka dwellers) of first generation		Recent migrant		Total sample (taking into account those whose residence period in Norilsk and Dudinka is unknown)	
	People	%	People	%	People	%	People	%
			Zheleznogorsk	1.3			Astrakhan	1.6
			Kaluga	1.3			Abakan	1.3
			Krasnodar	1.3			Ekaterinburg	1.3
			Norilsk	1.3			Ufa	1.3
			Ekaterinburg	1.3				
Migration to the city where one of the parents had been born	4	5.7	6	3.8	12	19.4	22	7.2

Source: Nadezhda Zamyatina and Aleksey Yashunsky.

However, in reality, there are few cases of such return migration (see Table 4.2).

Thus, in one degree or another, only about a tenth of graduates from Norilsk and Dudinka return to their "family network." The return to the hometown of one's parents is traced in even fewer cases, as the move "to the grandmother's residence" in about half of the cases is not linked with a return to the "historical homeland." Many grandmothers of the Northerners, for example, live in newly purchased apartments in Moscow or the greater Moscow region.

Another interesting result is that in many cases, there is an "interfamily" return not directly to the parents' place of origin, but to a major residential center in the same region. As a result, large regional centers of the European part of Russia and the Urals are settled by youth from families whose roots originate in the periphery of the respective centers. Natives of the small towns of the Krasnoyarsk krai, for example, spent a period of time in the Arctic before moving to Krasnoyarsk; natives of Chuvash and Mari towns and cities move to Nizhny Novgorod, and so on.

Thus, the interfamily "South–North–South" path is not circular, but spiral. The natives of small peripheral towns, after spending some time living in the North, send their children to their home regions, not directly to the native city, but to a more prestigious neighboring town.

The radius of geographical proximity between the homeland and the place of final residence may be as much as several hours of travel. As one respondent explained, "Yes, we live close: only three hours and we are there."[6] As the interviews reveal, this distance allows at most weekly face-to-face interaction with the relatives who remain at the original place of exodus (see Table 4.3).

Instead, communication (temporal proximity) among family members must be maintained within the framework of social networks. For example, one informant (originally from Western Ukraine) reported that her son's educational institution in Orel is convenient for her, because she stops to see her Ukrainian relatives on her way to visit her son. Likewise, she said that it is convenient for the natives of Eastern Ukraine to have a "base" in Belgorod when visiting their relatives.

The trends of returning to "family networks" – including spiral ones, which implies that a family member has transferred his/her residence to a more prestigious city – can be traced via the online social network VKontakte, by comparing them with the history of urban development and settlement in many cities of the North (see Table 4.4 and Figure 4.3). Return migration in West Siberian oil towns is especially noticeable. Here, for example, there is a traditionally large community of Tatars and Bashkirs;[7] therefore, many people from Noyabr'sk and Gubkinsky (Yamal-Nenets) go to Kazan and Ufa (the capitals of Tatarstan and Bashkortostan, respectively). There are other patterns of "reverse flow" based on family history. For example, the construction of the railway in the town of Gubkinsky in the 1980s was led by the Komsomol's *Molodogvardeitsi* squad from contemporary Belarus, whose leader, V.V. Lebedevich, became the city's mayor. As Table 4.4 shows, the flow of young people from Gubkinsky to Minsk, the capital of Belarus, is still significant.

Table 4.2 Expected level of "interfamily" return migration of Norilsk and Dudinka

Stated a move to the:	*Hometown of mother or father*	*Home region of mother or father*	*City (settlement) of grandmother's residence*	*Region of grandmother's residence*
The total number of people who stated a will to move	22	33	36	42
Total share of the respondents who mentioned this direction for migration (%)	7.2	10.7	11.7	13.7

Source: Nadezhda Zamyatina and Aleksey Yashunsky.

Table 4.3 Examples of "gravity zones" in action in major regional centers within the framework of interfamily migratory cycles "South–North–South" (examples revealed by the survey)

City of potential settlement of young migrants from Norilsk and Dudinka	Cities or regions from where his parents went to the North (excluding the cities of the same republic or the same area as the city of potential settlement)
Voronezh	Cities of Stary Oskol, Kursk, Volgodonsk
Ekaterinburg	Altai krai, Novosibirsk
Kazan	Mari El (city of Volzhsk), Chuvashia (city of Kanash)
Nizhniy Novgorod	Chuvashia (city of Kanash)
Novosibirsk	Chelyabinsk, Kemerovo, Irkutsk oblast (New Igirma settlement)
Omsk	Kazakhstan (city of Petropavlovsk)
Samara	Cities of Volgograd, Balakovo[1]
Ufa	Cities of Nefteyugansk, Asha

1. The questionnaire contained an explanation: "my relatives live near the city of Samara."
Source: Nadezhda Zamyatina and Aleksey Yashunsky.

Table 4.4 The most popular destinations for youth migrating from some Northern cities (no less than 0.5 percent of the total number of primary school graduates) according to the online social network VKontakte

Noyabrsk		Gubkinskiy		Magadan		Bratsk		Ust-Ilimsk	
Direction	Share of total (%)	Direction	Share of total (%)	Direction	Share of total (%)	Direction	Share of total (%)	Direction	Share of total (%)
Tyumen	6.3	Tyumen	8.6	St. Petersburg	6.6	Irkutsk	5.8	Irkutsk	9.8
St. Petersburg	3.5	Moscow	6.7	Moscow	6.2	Moscow	3.4	Novosibirsk	9.7
Moscow	3.3	St. Petersburg	3.7	Novosibirsk	1.3	St. Petersburg	3.1	Krasnoyarsk	5.8
Ekaterinburg	2.4	Ekaterinburg	2.9	Khabarovsk	1.2	Novosibirsk	3.0	Moscow	3.1
Novosibirsk	1.6	Ufa	2.1	Belgorod	0.9	Krasnoyarsk	2.6	St. Petersburg	2.2
Ufa	1.3	Omsk	1.6	Krasnodar	0.9	Krasnodar	0.5	Bratsk	1.4
Chelyabinsk	0.9	Krasnodar	1.2	Vladivostok	0.7	Tomsk	0.5	Ulan-Ude	0.7
Samara	0.9	Belgorod	1.0	Voronezh	0.6			Omsk	0.5
Omsk	0.8	Kiev	0.9	Irkutsk	0.5				
Krasnodar	0.7	Samara	0.9	Rostov-on-Don	0.5				
Belgorod	0.5	Chelyabinsk	0.8						

(continued)

Noyabrsk		Gubkinskiy		Magadan		Bratsk		Ust-Ilimsk	
Direction	Share of total (%)	Direction	Share of total (%)	Direction	Share of total (%)	Direction	Share of total (%)	Direction	Share of total (%)
Kazan	0.5	Kazan	0.7						
		Volgograd	0.6						
		Minsk	0.6						
		Novosibirsk	0.6						
		Noyabrsk	0.6						
		Perm	0.6						
		Purpe	0.5						
		Kurgan	0.5						
Population of the city (2013)	108.1	Population of the city (2013)	26.3	Population of the city (2013)	95.1	Population of the city (2013)	241.3	Population of the city (2013)	84.3
Total number of questionnaires	10756	Total number of questionnaires	1284	Total number of questionnaires	12739	Total number of questionnaires	19096	Total number of questionnaires	8418
% remaining in Noyabrsk	42.8	% remaining in Gubkinskoe	49.1	% remaining in Magadan	47.6	% remaining in Bratsk	73.9	% remaining in Ust-Ilimsk	56.0

Note: In every explored city, around 20 to 25 percent of the questionnaire cards did not contain any information about the place of migration. Therefore, all of the shares of the total number of investigated questionnaire cards should be regarded as minimal. For example, if the specified number of those remaining in Bratsk after graduating from school is 73.9 percent, this means that no less than 73.9 percent will remain.

Source: Nadezhda Zamyatina and Aleksey Yashunsky.

Migration in the capital: a departure from the network or a new network?

Migration to large regional centers can be defined as a specific "spiral" movement that occurs "over a generation" and is guided by social networks. At the first glance, migration to the major cities of the country can be interpreted as the result of a deliberate separation from a social network localized in one region or another. However, this is not always the case. As shown in Table 4.5, a third of Norilsk dwellers have someone to turn to for help to in Moscow and in St. Petersburg.

However, migration to Moscow, St. Petersburg, or regional capitals is entirely the result of social networks or social proximity. The institutional linkages between cities also play an important role. For example, branches of universities often channel young people to specific large cities (according to our data and Volosova 2008). This shows what important influence in shaping the key streams of young migrants is played by the links between central universities and their satellite campuses. Cities with large apartment blocs that were centrally built for Northern pensioners during the Soviet era also are attractive for the new Northerners, who often prefer to settle close to former neighbors and friends who shared life in the North.

In general, the qualitative interviews and questionnaires revealed interesting trends about migration flows of young people from the North. However, they do not allow for an accurate calculation of the role of social networks in the trajectory of migration flows in relation to migration to large cities. Therefore, we conducted a quantitative study using data from the online social network VKontakte.

What is more important in choosing a migration destination: social network or economic opportunity? In order to categorize the attraction of large cities as such (for example, due to economic factors, such as larger labor markets and other rational reasons not related to organized proximity) and their attraction within the framework of certain types of proximity, we carried out a regression analysis of the data collected through VKontakte. A model was constructed in which the value of migration flows between the two localities was determined by the ratio, similar to the law of universal gravitation; with this in mind, the dependent variable "mass" is replaced by the population size of settlements. Denoting the number of migrants as MIG, population size of settlements as $POP1$ and $POP2$, and the distance as D, we have the following:

$$MIG = POP1 \times POP2/D^2 \quad (5.1)$$

Accounting for some additional factors revises the formula as follows:

$$MIG = K_1^{a1} \times K_2^{a2} \times \ldots K_m^{am} \quad (5.2)$$

where K_i represents factor values, and ai represents exponents of levels. The ai indicators can be both positive and negative. Using the logarithm of the magnitude (e.g., decimal) in this relationship transforms the equation into a linear relationship:

$$\lg(MIG) = a1 \lg(K_1) + a2 \lg(K_2) + \ldots + am \lg(K_m) \quad (5.3)$$

Table 4.5 "Own" cities and social capital (cities with the largest number of positive ratings): high school students of Norilsk (including Dudinka)

City	Stated ability to obtain assistance of various kinds in this city (listed as number of people)							Own. cities close in spirit	Alien cities
	Temporary housing	Job	Financial assistance	Business	University education	Vacation	Mean		
Krasnoyarsk	77	46	38	22	43	35	43.5	117	44
St. Petersburg	71	45	26	25	43	36	41.0	106	68
Moscow	83	59	38	25	37	27	44.8	87	93
Dudinka	3	5	7	2	3	11	5.2	75	45
Talnah	2	0	0	0	0	1	0.5	54	46
London	1	0	0	0	0	0	0.2	43	127
Abakan	11	6	5	1	5	5	5.5	38	87
Kayerkan	2	0	1	0	0	1	0.7	38	49
Krasnodar	11	6	8	3	9	4	6.8	35	77
Novosibirsk	18	3	8	6	8	8	8.5	31	66
Ekaterinburg	5	2	1	1	2	1	2.2	27	84
Kazan	6	4	2	1	2	8	3.8	23	97

(continued)

Table 4.5 "Own" cities and social capital (cities with the largest number of positive ratings): high school students of Norilsk (including Dudinka) (continued)

City	Stated ability to obtain assistance of various kinds in this city (listed as number of people)							Own. cities close in spirit	Alien cities
	Temporary housing	Job	Financial assistance	Business	University education	Vacation	Mean		
Nizhniy Novgorod	6	4	4	2	4	3	3.8	23	84
Belgorod	5	3	4	1	6	2	3.5	22	94
Omsk	3	3	3	3	5	4	3.5	21	92
Rostov-on-Don	6	3	3	0	3	5	3.3	19	91
Baku	8	5	7	5	7	7	6.5	18	147
Correlation coefficient with the number of those who indicated the respective cities as "their own" (entire sample)	0.84	0.83	0.84	0.83	0.85	0.88	–	–	–

If there is a sufficient number of migratory flows, subject to the same laws, then with the known values of migration flows *MIG* and factors *K1*, …, *Km*, the coefficients "*a1*", …, "*am*" can be defined by a linear regression.

In order to avoid additional errors in the model, we selected streams with at least three migrants in a sample. The purpose of mathematical modeling in this case is not so much the definition of the importance of various factors ("*ai*" coefficients), as the search for settlements, standing out from the overall sample. It is assumed that the model describes the rational reasons for choosing migration patterns, outside of the logic of organized types of proximity; the outliers are cities that attract migrants with these or other types of organized proximity.

Sets of outlier cities, obtained through this manner, can be seen as cities that require further study. The initial data used migration flows from Igarka, Dudinka, and Norilsk into the major cities of the Russian Federation. The following regression equation was used:

Equation (5.3) was used with the following factors:

1. POPUL Population size of the target city (in number of people, according to the Federal State Statistical Service data for 2013)
2. DIST Distance from the city of origin to the target city (great circle)
3. REG Whether the city of origin and the target city belong to the same subject of the Federation (dummy variable equals 1 if from the same subject of the Federation, and 0 otherwise)
4. CAPITAL An indicator that the target city is the capital (a dummy variable, equal to 1 if the target city is Moscow, and 0 otherwise)
5. CENTER An indicator that the target city is the capital of a federal subject (a dummy variable equal to 1 if the target city is the center of the federal subject, and 0 in other cases)
6. PRICE M2 Average cost per square meter in the primary housing market in the region of the target city (expressed in Russian rubles, according to the State Statistical Committee data for 2011)
7. W The average monthly wage (calculated by weighting the average gross salary in individual sectors of the economy on the number of employees in the organizations of the relevant industry, for 2010–2012, according to the Federal State Statistical Service)

When switching back from the logarithms to the multiplicative form of representation, the formula becomes:

$$MIG = 10^{a0} \cdot POPUL^{a1} \cdot DIST^{a2} \cdot 10^{a3.REG} \cdot 10^{a4.CAPITAL} \cdot 10^{a5.CENTER} \cdot PRICE\ M2^{a6} \cdot W^{a7} \quad (5.4)$$

The resulting linear regression values for each of the factors are listed in the chart title of the corresponding city of origin for migration. On the basis of the parameters obtained in such a manner, a "predicted" value was found for the number of migrants in each of the target cities. Each point of the graphs shows the ratio of

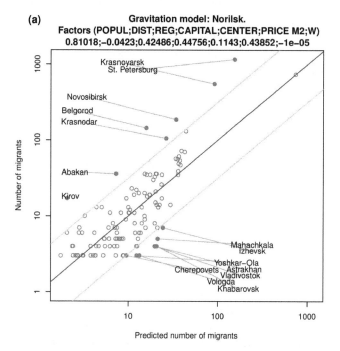

(a)

Gravitation model: Norilsk.
Factors (POPUL;DIST;REG;CAPITAL;CENTER;PRICE M2;W)
0.81018;–0.0423;0.42486;0.44756;0.1143;0.43852;–1e–05

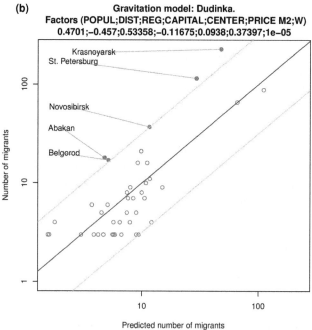

(b)

Gravitation model: Dudinka.
Factors (POPUL;DIST;REG;CAPITAL;CENTER;PRICE M2;W)
0.4701;–0.457;0.53358;–0.11675;0.0938;0.37397;1e–05

Figure 4.1 (a) Predicted number of migrants for Norilsk. (b) Predicted number of migrants for Dudinka.

Source: Nadezhda Zamyatina and Aleksey Yashunsky.

"forecasted" number and the actual number of migrants in the given sample. The scale on each axis is logarithmic.

We chose a difference of 0.5 between the logarithms of the forecast and the actual value of migration flows as a threshold for the definition of "non-standard" goals. That is, if the flows differed $10^{0.5} \approx 3$ times, it was believed that a target city does not fit into the model. Based on the constructed graphs, a list of non-standard targets can be distinguished.

Figure 4.1 shows that the population of the city of incoming migration, its administrative status, and position in the same subject of the Russian Federation with the city of outbound migration play a key role in shaping the flow of migration from the studied Northern cities. The first two factors can be considered universal indicators of attraction for a Russian city: population size determines the capacity of the labor market, while administrative status – in the Russian context – indicates the comfort of social conditions and the availability of a variety of facilities and prestigious universities, the last factor being especially important in the context of youth migration.

It is a trend that the city's average wage level has almost no effect on the choice of direction of migration – probably because our study primarily deals with educational migration. As the interviews revealed, underestimating the labor market in the city of potential migration often leads to the need for secondary migration, and even a return to the North.

As shown in the regression analysis, Moscow is one of the most popular migration destinations, a finding well within the "economic-administrative" logic and regardless of social proximity. Migration in Krasnoyarsk is also evidently associated with the capital status of the city in the Krasnoyarsk region, where the studied cities are located. However, the increased flow of migration to St. Petersburg, Novosibirsk, Abakan, and Belgorod demanded further explanations. As a result, factors more complex than the "simple" social proximity of attraction were identified, although still associated with it. We will analyze the factors of attraction for St. Petersburg in more detail.

If not social networks, then what? "Nonfamily" factors attracting young people from the Arctic

The appeal of St. Petersburg can be summed up in two words: "our own." For the majority of Norilsk and Dudinka residents, St. Petersburg is considered to be "their" city in the mental aspect predicted by the distribution of social capital (Table 4.6). A detailed analysis of the questionnaires revealed that St. Petersburg is sometimes called "our own" even by those who have no one in the city to help them. In contrast, Moscow is often regarded as unpleasant even by individuals who have close relatives and friends in Moscow.

Furthermore, the volume of social capital ascribed to St. Petersburg increases according to the length of stay in Norilsk (Dudinka) (see Table 4.7). The longer a family stays in the North, the more it becomes associated with St. Petersburg.

Table 4.6 Comparative preferences of Moscow and St. Petersburg

Respondents	St. Petersburg	Moscow	Both are close	Both are unpleasant	Have never thought about it	Indifferent to both
Total number of people	111	68	25	31	62	1
Total %	36.2	22.1	8.1	10.1	20.2	0.3
Natives, number of people	26	15	6	7	13	1
Locals, %	37.1	21.4	8.6	10.0	18.6	1.4
First genera-tion, number of people	60	38	10	18	28	0
First genera-tion, %	38.0	24.1	6.3	11.4	17.7	0.0
Migrants, number of people	18	13	9	1	18	0
Migrants, %	29.0	21.0	14.5	1.6	29.0	0.0

Table 4.7 Social capital of Norilsk and Dudinka dwellers in different social groups (based on the time of residence in the North): share indicating they "have someone to live with for a few days" in this city

Respondents (%)	St. Petersburg	Moscow	Krasnoyarsk	Novosibirsk	Abakan
Natives	28.6	25.7	34.3	7.1	4.3
First generation	23.4	31.6	23.4	5.7	1.9
Migrants	8.2	21.3	21.3	4.9	6.6

Source: Nadezhda Zamyatina and Aleksey Yashunsky.

Social capital in Moscow among Northerners of different generations remains remarkably steady. At least one-fifth of the young Northerners have someone they could live with in the Russian capital. Some 20–25 percent of the young Northerners consider Moscow to be "their" city (slightly less than the number of individuals who have close relatives and acquaintances in Moscow). But 37–38 percent of the first- and second-generation Northerners consider St. Petersburg to be "theirs" – a significantly higher level than those who have someone to turn to for help in that city. A much smaller proportion of respondents among recent migrants consider St. Petersburg to be "theirs."

Thus, there are some factors that generate a positive attitude toward St. Petersburg during a long stay in Norilsk. Resorting to metaphors from the arsenal of the economists and paraphrasing Marshall (1920), we can say that love for St. Petersburg is "diffused in the atmosphere of Norilsk."

This conclusion is confirmed by another conclusion drawn from our data. We can compare attitudes toward St. Petersburg and Moscow among those individuals who are definitely going to a third city, such as Krasnoyarsk, regional centers, and others (see Table 4.8).

Table 4.8 Variations of mental space (peaks of attractiveness) and migration preferences

Groups of potential migrants to:	*Percentage of responses to the question: What city is closer to you: Moscow or St. Petersburg?*					
	St. Petersburg	*Moscow*	*Both are close*	*Both are unattractive*	*Have not thought about it*	*Indifferent to both*
St Petersburg	81.6	5.3	10.5	0.0	0.0	0.0
Moscow	0.0	63.2	21.1	10.5	5.3	0.0
Krasnoyarsk	37.8	13.5	5.4	10.8	27.0	0.0
Large regional centers, aside from Krasnoyarsk[1]	36.7	30.0	3.3	20.0	10.0	0.0
Secondary regional centers outside of Central Russia[2]	27.3	31.8	9.1	9.1	31.8	0.0
Small towns in the south of Krasnoyarsk krai and Khakassia	42.9	14.3	0.0	14.3	42.9	0.0
Azerbaijan	0.0	42.9	28.6	14.3	14.3	0.0

Notes
1. Novosibirsk, Nizhny Novgorod, Kazan, Ufa, Ekaterinburg, Omsk, as well as (single cases) Samara and Saratov.
2. Astrakhan, Belgorod, Cheboksary, Yoshkar-Ola, Tambov, Penza, Ulyanovsk, Kaliningrad, Barnaul, Vladikavkaz, Krasnodar, and Stavropol.

The Krasnoyarsk residents likely to prefer St. Petersburg are mostly native-born Northerners, or, in any case, Siberians – immigrants from the southern portion of the Krasnoyarsk krai (Kansk, Bogotol, etc.). Those who arrived in Norilsk from Azerbaijan clearly prefer Moscow; those who came from the European part of Russia also prefer Moscow. Thus, it can be argued that the preference of St. Petersburg is the Norilsk, or perhaps somewhat more broad, Northern or Siberian *specificity*.[8]

What are the roots of these preferences for St. Petersburg? Norilsk residents' preference for St. Petersburg relates to the huge role that Leningrad residents[9] have played in the history of Norilsk. According to V.V. Larin, a biologist and patriot from Taimyr, now working at a museum of the history and development of the Norilsk industrial region:

> First of all – 1953. Leningrad contingent was largely presented here [Case of Comrade Kirov[10]]. Second, the Institute of Agriculture of the Far North. Its fate is as follows: N.S. Khrushchev, in addition to growing the King of the Fields [corn] … had another idea: "Fellow scientists, let's get closer to production!" Well, and physicists were sent to Dubna, while the Institute for Polar Agriculture was luckier: "Let's go, pack your bags and move out to Norilsk!" It was 55 years ago; it just celebrated his 55th birthday. [...] Third, Norilsk industrial district and the city of Norilsk – the most powerful nest, one of the most powerful geological schools of the USSR. The Russian Geological Research Institute (VSEGEI) – they have a branch here. And here was a powerful stream of unfading St. Petersburg geologists for many, many years.

At the same time, some experts, such as the local journalists and historians Stanislav and Larisa Stryuchkov, told us that the historical connection between Norilsk and Leningrad is exaggerated, that even the famous Leningrad influence in Norilsk architecture was very limited. The first chief architect of the city of Norilsk, V. Nepokoichitsky, was born and studied in Leningrad before he was invited to a position in Norilsk at the age of 29. Many prominent architects, including some more experienced than Nepokoichitsky, had been caught in Stalin's purges and were prisoners assigned to work for him. Some of them managed to "smuggle" into the face of Norilsk many features of Armenian (M.D. Mazmanian, G.B. Kochar, and others)[11] and Baltic (Latvian Y.K. Trushinsh, etc.) architecture.[12] Apparently, while the relationship of Norilsk with Leningrad/St. Petersburg has genuine roots, it has become mythologized and self-reinforcing in the Norilsk community.

Residents' traditional gravitation toward St. Petersburg, in turn, prompted real estate agents to establish branch offices in Norilsk and to offer special deals on real estate in St. Petersburg (see Figure 4.2). Thus St. Petersburg gained a secondary organizational proximity to Norilsk:

Now the Lenspeksmu construction firm provides a discount if the child enters a university in St. Petersburg.[13]

(continued)

Figure 4.2 Advertising for buying apartments in St. Petersburg, Moscow, or Belgorod in Norilsk's streets (photos by the authors).

As a result, there is a "chain reaction" of self-sustaining migration from Norilsk to St. Petersburg that endures, even when the return flow of Leningrad natives from Norilsk is apparently "exhausted" (the respondents often lamented that all Leningrad natives have left the area).

Similarly, institutional ties with St. Petersburg are manifested primarily in the desire of Norilsk residents to send their children to the more prestigious universities in St. Petersburg:

> Last year we had 53 graduates. All of them entered a university, including 22 people who received state scholarships to study in Ekaterinburg; Nizhny Novgorod – there is a good linguistic university, and Krasnoyarsk (SFU); St. Petersburg – just solid. Two of my daughters studied there. In the past they had targeted intake into the INZHEKO - Engineering and Economic University (back then they were held the 6th place in Russia based on the quality of knowledge). The University of Trade Unions had an agreement – they would come, conduct an exam, there was no USE [Unified State Examination] at our base. Among the departments there were social pedagogy, and sound engineering (at the theater department). Graduates of such departments face employment problems here, so they stayed there [in St. Petersburg where they attended university]. INZHEKO also conducted the entrance test here and taught mathematics (they entered into an agreement with the Scientific Research Institute, and their professors taught the kids here).[14]

The current positive image of St. Petersburg is perhaps partly attributable to the transformation of ideas about it being cheaper than living in Moscow. Affordability joined availability and proximity as reasons to move to the former capital city:

> First of all, the people there are normal, housing prices are lower, it is a quiet town, compared with Moscow. Don't get me wrong, but Moscow is a crazy city. Everything there is tied around money, huge money. A policeman can stop a student on the street and arrest him/her, and arrest is the best-case scenario.
>
> And in St. Petersburg there are normal people. There are still children of those who survived the blockade. If you ask for directions, they always stop, explain everything.[15]

However, a positive image of St. Petersburg would hardly have been shaped only on the basis of price preferences. We believe that this is just one of the modern aspects of an exceptionally positive image of St. Petersburg in the Russian culture in general, and in Norilsk in particular.

At the heart of the Norilsk myth of St. Petersburg is social proximity, coupled with a large number of Leningrad natives involved in the early history of formation of Norilsk.[16] This social proximity gave rise to a secondary, organizational proximity (branches of universities, policy of realtors). However, there is yet another form of proximity, one associated with the high degree of affinity with certain geographic features based on information, local myths, and culture. We call this a "mental proximity." It is also fueled by personal trips (temporal geographic proximity).

As a result, some geographical objects appear closer and more attractive than others – solely due to the established image. This is especially true for the more distant objects. A good example is the positive image of London (see Table 4.5), which for many Norilsk residents is "closer" than the neighboring satellite town of Kayerkan.

We tend to believe that this is due to the specific conditions of the North: remoteness and isolation from the main zone of settlement makes all possible contacts between the North and the "mainland" more complicated. Thus individuals tend to retrace the earlier "trodden" path to reduce the high cost of interaction. Social ties are reinforced for long periods of time, due to the multiplier superstructure organizational and institutional proximity, as well as mental proximity.

The mental proximity of St. Petersburg, however, does not automatically supersede social and institutional proximity, which clearly shows the negative perception of Moscow compared to St. Petersburg. These attitudes have been formed by specific local cultural processes and narratives.[17] In general, the type of mental proximity can be characterized as an intuitive preference (in the course of certain actions) of some geographic features over others. At the basis of differences in preferences, we tend to see the differences in the evaluation of various places of symbolic capital (transferring Bourdieu's concept of symbolic capital from the individual to the geographical object).

Similar processes have caused increased migration flows that did not fit into the general trend identified in the course of regression analysis, for example, migration to Belgorod. According to our informants, the key role here is played by the city's location near the border with Ukraine, which makes it attractive for Ukrainians who want to live close to their relatives in Ukraine, while remaining in Russia. It also played an important role in organizational proximity: in the 1990s many large-scale housing projects were completed for Northern pensioners who had been collectively resettled in Belgorod. Today, specialized real estate companies are conducting similar projects, with the encouragement of local authorities, which is strengthening the existing trends. Finally, today Belgorod enjoys a good reputation among Northerners, and Belgorod University is very familiar to them.

Another popular destination is Novosibirsk, home to one of Russia's leading universities – Novosibirsk State University (NGU). However, our surveys show that migration to the city is related more to social networks than the status and appeal of NGU. In particular, "Novosibirsk residents" of Dudinka, graduates of the Institute of Water Transport:

Important migration flows: Odessa, St. Petersburg, and Moscow "schools." Odessa – University of Water Transport, the Port gathered high-class specialists, there is still a lot of experts in Odessa. In Novosibirsk, there is also a University of Water Transport; people were recruited from there as well. And then there has been "established a clan" – parents train their children in the same specialty, and now their children work there.[18]

"North diaspora": new networks of social proximity

Characteristically, the presence of friends in Moscow among the potential migrants is minimal – and the same variable is maximal among the potential migrants to the peripheral centers and Azerbaijan (see Table 4.9). Using our survey on the intentions of high schoolers, we could summarize that people who believe in their ability to create new social relationships migrate to big cities, while those who rely on existing "old" social relations (these are strong social ties, according to M. Granovetter) prefer the peripheral cities.

The interviews also show that in the course of a long stay in Norilsk, Norilsk dwellers form a new circle of friends, resulting in the fact that during migration, such families rely not so much on the social networks of relatives as on the new social connections (one of the informants emphasized the fact that they migrate more to be closer to friends than relatives).

Table 4.9 Variation in the role of strong social ties in the city of potential migration in the choice of migration destination

Groups of potential migrants	Proportion of group members who have relatives and close friends in the city of potential migration (%)
To Moscow	57.9
To St. Petersburg	78.9
To Krasnoyarsk	81.1
To large regional centers	78.6
To the secondary regional centers outside of Central Russia	76.0
To the south of Krasnoyarsk krai	77.8
To Azerbaijan	100.0

Source: Nadezhda Zamyatina and Aleksey Yashunsky.

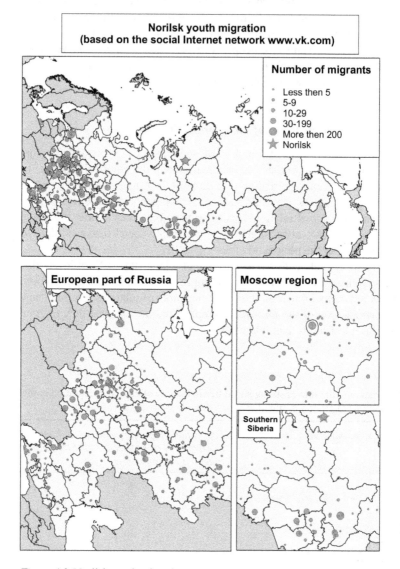

Figure 4.3 Norilsk youth migration.

Source: Nadezhda Zamyatina and Aleksey Yashunsky.

Conclusion

The primary hypothesis that the current migration from the North is configured by social networks and is the final multigenerational family driving cycle of "South–North–South" was confirmed only partially. Only about one-tenth of young people leaving the North are guided by their social capital when choosing the destination city. This type of behavior is primarily observed among recent migrants to the North, who, after a period of time spent in the Arctic, return to their former place of residence.

In this case, we found not only circular trajectories, which we expected, but also the spiral trajectories of family movement "South–North–South." These patterns are typical of families whose members initially went to the Arctic from small peripheral cities and towns. After a period of residence in the North they themselves (and/or their children) are not returning directly to their home town/village, but to a larger and more prestigious city in the same region. Thus, the cycle of "passage" through the North and the consequent return to the native region raises a family's symbolic capital and advances family members up the social ladder, which is expressed (among other ways) through a change of residence.

A new fact discovered in the study is the "fouling" of social networks with other types of proximity: the lines of social ties are repeatedly "thickened" through institutional and organizational links, as well as specific types of mental proximity. Mental proximity is expressed in the intuitive preference for some geographical features over others, which are based, presumably, on the evaluation of differences in the symbolic capital of different locations.

Notes

1 Research for this chapter was funded in part by a grant from the Russian Science Foundation (Project No. 14-38-00031, "Foundation of Laboratory of Complex Geocultural Studies of the Arctic").
2 A June 2014 article in the Yamal press noted: "Not everyone prefers the nearby territory. Many headed to the Far North, including Yamal, and in particular Noyabr'sk – they share blood ties and friendship with our region. To mention some of the typical jokes that live to this day, since the great development of the Far North, people call these areas – 'Yamal-Donetsk district' and 'Lugansk tundra.' In other words, this area has always been populated by a lot of immigrants from Ukraine. Many came for half a year, but remained forever. Thus, in Noyabr'sk, according to the latest census, over 40 percent of the population are ethnic Ukrainians. ... Many Mariupol dwellers, including Alexander and his wife, were forced to leave home and go on the run, due to the above mentioned events. The wife of our hero remained in Moscow, where her daughter lives with her. Alexander himself went to seek support and refuge here in Noyabr'sk, where his good friend has been living for many years now." (Anisimova 2014)
3 In this context, they often discuss the theme of "death of geography."
4 In the 1960s, one expert on northward migration in the USSR explained the exodus of the older generation from the North by citing labor migrant's preservation of ties with their original places of origin. Although the term "social networks" was not used back then, it is precisely what is meant: "By far not all consider the North as a permanent residence – there are few of those who do. Each year, tens of thousands of people who have lived in the North for a decade or more, leave its districts for good. The new generation of residents settles in the abandoned, but not yet cooled, seats of the old-timers who have left the North. Is this a consistent pattern? We think so. We must not forget that the population in most Northern cities and industrial towns is formed only by the first generation of settlers, attracted from other parts of the country, and for a number of reasons, the relation of this population with the areas of origin is rather strong. This tendency will prevail in the subsequent generations. Therefore, the exodus of the older age groups of the Northern regions will, apparently, be the case for decades." (Yanovsky 1969, 37)

5 This chapter does not cover the material on Igarka in detail: in general, the conclusions on Igarka are similar to conclusions on Norilsk and Dudinka, but with a smaller geographic range of potential migration; almost all of the respondents are planning to move to Krasnoyarsk.

6 From an interview with an online-learning student (studying in Krasnoyarsk).

7 The Middle Volga is an older oil-producing area that recruited experienced staff for oil and gas extraction projects.

8 There is also evidence of a clear preference for St. Petersburg/Leningrad in Igarka during the Soviet times: "Of course, not only Leningrad residents were exiled to the Yenisei and recruited, and yet they were not lost in a motley mixture of multinational entities, and the fate of political cases, which the new port on the Yenisei became from the first years of its existence. Not the least of the reasons for this Leningrad identity can be considered the high professionalism and loyalty to the traditions of the St. Petersburg culture. But, probably, this alone would not be enough for the very respectful attitude to the word *Leningradets* [native of Leningrad] to down from generation to generation in the Arctic. Igarka's Germans, Greeks, Lithuanians, Finns, too, cannot be accused of betraying the traditions of the national culture, and great experts in their fields have always existed among the many Siberians, people from Arkhangelsk, or, say, the Muscovites who lived in Igarka. As I see it, the key to understanding the causes of the special relationship of Igarka residents to Leningrad residents was given in the words of the head of the Igarka KGB, Kurbatov, which he pronounced to his subordinates in the autumn of 1938: 'Until we completely clean away the backbone of Leningrad natives, our work in Igarka will be pointless. Either we eliminate to hell their disobedience, or we will lose the city'" (Gorchakov 2015).

9 Between 1924 and 1991, the city of St. Petersburg was called Leningrad.

10 Many of the Leningraders convicted for conspiring to kill Leningrad Communist Party chief Sergei Kirov in 1934 were sent to Norilsk.

11 See selection of photos on the Norilsk History Blog, http://severok1979.livejournal.com/47732.html.

12 See the list of names of architects imprisoned in Norilsk, specifying their educational background (Slabukha 2010). See also the Repressed Architects Memorial Database: http://www.memorial.krsk.ru/DOKUMENT/People/Arhitek.htm.

13 Author's interview with an employee of the Department of Education of the City of Norilsk.

14 Author's interview with the head teacher of a school in Norilsk.

15 Author's interview with a Norilsk resident, about 50 years of age.

16 There were several Komsomol members from Leningrad, Moscow, and a few other cities who worked in Norilsk in 1956. In addition, the Leningrad universities were the traditional training ground for specialists on the Northern culture and economy. As a result, the Russian Arctic brimmed with people who were born and/or studied in Leningrad.

17 In particular, Irkutsk sociologist Michael Rozhansky drew our attention to the fact that, in many Soviet films about Siberia dated from the 1950s–1970s, the villains tend to be Muscovites, who embodied the image of cynical materialism in contrast to the openhearted Siberians.

18 Author's interview with an employee of the Dudinka city administration.

5 Infinite travel

The impact of labor conditions on mobility potential in the Northern Russian petroleum industry[1]

Gertrude Saxinger

Workers in the Northern Russian oil and gas industry often describe their lives as "extreme." The adjective refers to working in a physically challenging environment – be it the harsh climate, the hard industrial work, or the long commute to and from work. It also reflects the psychological pressure of living together with colleagues in small closed camps, the lack of privacy, and the hardship of being away from family for long periods of time. Those who have already spent decades on long-distance commuting (LDC), also called fly-in/fly-out work (FIFO), are now accustomed to this lifestyle; it has become their new "normal." In contrast, many new workers who have just started the arduous routine doubt whether this is a sustainable way to earn a livelihood and quit after a few years – or even a few shifts. Not everyone has the adventurous spirit and mental toughness to commute long-term.

This chapter explores how the working conditions in the petroleum industry and the living conditions in the camps in the Russian Arctic and sub-Arctic regions affect the mobility potential of construction and operation workers in the oil and gas sector. It seeks to identify what circumstances enable individuals to embrace a long-term, sustainable career in this difficult environment.

Understanding the motivations for and objections to mobility and to a life lived across multiple localities is crucial for, first, an industry that has a high demand for a mobile, skilled workforce and, second, for people considering pursuing such work. Especially for people from socioeconomically disadvantaged regions in central and southern Russia, there is an advantage in being ready to accept a mobile job in order to escape local unemployment and achieve social mobility (Eilmsteiner-Saxinger 2011, 2013; Saxinger 2016a, 2016b; Saxinger et al. 2014; Saxinger et al. 2015). The "fly-over effect" (Storey 2001) describes the usually negative dynamic whereby the salaries of LDC workers flow directly to the regions of residence, bypassing the region where the work is carried out. In the Russian case, however, this fly-over of revenues to southern and central regions is crucial since these areas are characterized by weaker economies (except in the regional capitals, Moscow and St. Petersburg) than that of the prospering Northern petroleum industrial regions. The spending power of these workers at home is vital for regional development (Öfner 2014; Saxinger 2016b; Saxinger et al. 2016).

This chapter offers three arguments. First, a mobile and multi-local life is characterized by the emotional and practical separation and connection of three key realms: at home, on duty, and journeying (Eilmsteiner-Saxinger 2010, Saxinger 2016b). The circumstances of each setting affect the other two realms. LDC work must be arranged in such a manner that people can negotiate these spheres in a balanced way, so that they can make an integrated sense of life.

Second, weak labor conditions and non-satisfactory living conditions in company camps affect one's private life in a way that frequently leads workers to drop out of this specific job market at an early stage. For them, long-distance commuting is not a sustainable way of making a living, and therefore they leave a sector that otherwise provides some of the highest-paid jobs for blue-collar workers as well as for engineers, administration, and management staff. By giving up high-paid jobs for lower wages – not to mention unemployment – workers may be deprived of a sustainable livelihood and socioeconomic mobility. The industry thus faces an ever-shrinking labor pool at a time when the need for highly qualified personnel in the remote regions is constantly increasing. However, the long-term consequences of the continuously dropping oil prices since 2014 are unknown – both internationally as well as in Russia.

Third, despite a steady increase in state control over natural resources and the respective industries in contemporary Russia, the state shows no interest in revisiting the existing laws regarding labor conditions or implementing legal reforms related to LDC in this sector. The labor market in the petroleum industry is left exposed to the global neoliberal dynamics of financial efficiency that can result in harsher working conditions. In this particular context, I speak of "re-socialist neoliberalism" in terms of state control over the natural resources and state neglect of employee well-being. As shown below, the degradation of labor conditions occurs primarily in the construction sector of the petroleum industry, which is characterized by outsourcing to smaller subcontracting companies.

The next section outlines my theoretical framework where the notion of the normality of an itinerant work-life will be discussed in the context of multi-locality and mobility. The methodological section introduces the region used for this case study and highlights the need for an ethnographic, qualitative approach to a field that is seldom examined in the social sciences and, if so, primarily from a quantitative perspective. Following this, I will describe the LDC system for labor force provision in remote construction and production. Finally, the travel and labor conditions under which LDC workers live and work are described along two poles of good to poor. The broad notion of "labor conditions" is divided into three components: employment conditions, living conditions in camps, and safety.

Mobility, multi-locality, and normality

Generally, LDC is defined as a form of labor where the workplace is such a distance from home that a daily return is not feasible. In some forms of LDC work, housing is provided at the workplace (for further definitions see Öhman and Lindgren 2003; Hobart 1979; Spies 2009; Storey 2001). The notion of "long" in LDC involves a whole set of meanings of distance: temporal, spatial-geographical, economic (e.g., affordability of transport home by workers or the willingness of the employer to pay for it), and technological distance (availability, speed, and comfort of transport).

LDC is considered to be a way of life involving hard work under extreme conditions (cf. Ananenkov et al. 2005; Andreyev et al. 2009; Bondarenko et al. 2003; Gareyev et al. 2002). The mobility of LDC workers is implicitly and explicitly defined as deviating from mobility norms like daily commuting, as well as from the prevailing form of sedentary careers in the industrial era. In addition, the extreme climate conditions in circumpolar regions as well as the demands for working under the open sky or hard industrial labor under dirty and dangerous conditions contribute to the perception that Arctic and sub-Arctic LDC work in the petroleum sector qualifies as extreme. Moreover, in some definitions the LDC way of life is considered to be bearable only as a last resort (Andreyev et al. 2009). However, this viewpoint neglects the normalization process in LDC life. It focuses on the problematic aspects of LDC, overlooking the ways in which LDC workers cope and make sense of this way of life. Another aspect of the "problematization" of LDC is reflected in the public discourse on LDC in the context of alcohol and drug abuse, prostitution, broken marriages, and other anti-social behavior and deviances. My approach, in contrast, highlights the ways of coping with difficulties and how people mitigate the extreme with normality, which can subsequently lead to the normalization of this mobile and multi-local lifestyle. Nevertheless, this study does not ignore the downsides of LDC (for more on prostitution and sexually transmitted diseases in the context of FIFO and LDC, see Saxinger and Nuykina 2015).

Although today the prevailing discourse states that we live in the age of mobility, which has replaced the sedentary age, mobility has always been a characteristic way of life of humans (Rolshoven 2006, 2008, 2009; Wolf 1982). Sedentarism is a rather new phenomenon and relates to nation-state building in Europe in the late modern age and the state's control over its citizens (Rolshoven 2009). At the same time, the age of industrialization brought new transportation technologies and relocated most work outside of the home, unlike the agrarian age (Weichhart 2009). This shift has led to increased mobility and multi-locality, which contributes also to today's increased transnational movements.

Today, the "normality" of mobility and multi-locality falls between two distant poles along a Gaussian distribution, if normality is considered to be agency and the ideal of the societal majority (Gerhard et al. 2003). While mobility is stigmatized in the case of Roma, vagabonds, and refugees, it is associated with "heroes" when applied to adventurers, extreme tourism, and academic and business mobility (Rolshoven 2011). Both notions are at play regarding mobile and multi-local LDC workers. They are considered unreliable people with an erratic lifestyle, yet they have prestige, since they are considered to be healthy, resilient, and ambitious because they handle this extreme work and cope with this type of life.

As my ethnographic data show, to outsiders, the LDC way of life – mobility and multi-locality – is not the norm because it clashes with the idea of sedentarism. It becomes clear that the construction and making of normality has a crucial role when it comes to coping with LDC. For those who already undertake LDC long-term, life with and between different places is successfully negotiated. Those who struggle with mobility and multi-locality and who cannot adapt to it quit their jobs after a few years or even a few shifts. They did not successfully master this negotiation of places. For them, LDC remains a non-normal lifestyle. They live, or must live, "two halves of a life," as some say. They see their life as non-normal, in the same way as the outside society does. In turn, those who have internalized this mobile and multi-local lifestyle say that they live a "doubled life" in the sense of added value. One interview partner pointed out: "The benefit is that I can live two lives in one life-span." This process of making normality has been described as "flexible normalism" (Link 1997, 425). Normality is not given by nature, but is a product created by the ideas, values, and agency of people. New centers of normality occur outside the normality of the societal average. It is a process of fluid borders and not of closed zones between deviance and normality. As Bourdieu shows, this process is also habituation in the social space and in the individual and collective locality of the actors (Bourdieu 1977, 1984). In this way, a socialization of practices occurs (Bublitz 2003, 151) and, at the same time, new and diverging normalities emerge.

Methodology

This study is based on anthropological fieldwork carried out between 2007 and 2010 in the Yamal-Nenets autonomous district (YNAO) and the Khanty-Mansi autonomous district–Yugra (KMAO), both in Western Siberia (see Map 5.1). Furthermore, I visited families of LDC workers in the central Russian republics of Bashkortostan, Chuvashia, and Tatarstan. A key mobile method was accompanying LDC workers on commuter trains between Moscow and Novy Urengoy in YNAO, the center for gas extraction in the Russian North. I made these three-and-a-half-day train journeys several times, travelling in total over 25,000 kilometers. I also had several stopovers in major towns like Surgut, Nizhnevartovsk, Pyt-Yakh, and Raduzhny in KMAO, the central regions for oil extraction. These travels on the train were superb opportunities to engage with workers while they were relaxed and had time to talk. In addition to statistical survey data, my field research is informed by observation, informal talks, narrative interviews with workers, their spouses, and children, and interviews with experts such as company representatives and scholars in this field. Grounded theory (Glaser and Strauss 1967) informed the analysis. While the majority of the – rare – scientific work on LDC management, health, and economic rationales is based on statistical data, this chapter draws on interpretative methodologies and contributes to a bottom-up view of the people involved in LDC.

Long-distance commute work

Since the 1980s and particularly in the last two decades, LDC has become increasingly important in the Russian North for supplying the workforce in the petroleum industry. The activities of resource exploitation, which are becoming ever more remote from (Northern) urban agglomerations (e.g., extraction sites on the Yamal Peninsula), require increasing mobility and consequently a multi-local lifestyle. The labor force supply in Northern industrial cities can no longer meet the increasing demand for workers. LDC is more feasible than expanding Northern towns to accommodate a substantial influx of new workers since the construction and maintenance costs of Northern towns is approximately four times higher than in more temperate regions (Hill and Gaddy 2003, Nuykina 2011). Furthermore, the LDC workforce is cheaper than local employees since the so-called Northern supplementary payments (*severnaya nadbavka*) must be paid only during the actual time workers are in residence in the North, not while they are off-duty at home in the south. In contrast, Northern-domiciled workers are entitled to these payments all year round. This makes it attractive for the industry to recruit LDC workers from the southern and central regions (Eilmsteiner-Saxinger 2011, Saxinger 2016b).

Mobility and multi-locality are the central characteristics of LDC work: a place at home for times off-duty, camp-life while on-duty, and considerable time spent journeying back and forth between these living and work spaces several times a year. Interregional LDC workers (*mezhregional'nye vakhtoviki*) travel, for example, over 3,000 kilometers from southern Russia and the North Caucasus, or over 1,500 kilometers from the Republic of Bashkortostan to Northwestern Siberia. The route from Moscow to Sakhalin requires a trip by airplane of some 9,000 kilometers and transits several time zones. Intra-regional LDC workers (*vnutregional'nye vakhtoviki*) still travel several hundreds of kilometers from their home in a Northern mono-industrial town to the actual worksite. These men and women are highly mobile and integrate physical, social, and emotionally distant spaces.

The interregionally commuting workforce from central and southern Russia is usually assigned a shift of 30 days onsite and 30 days at home (30/30). This time off is classified as a rest period, not as holiday. Common rotations also include 45/30 and 60/30. Since the global economic crisis of 2008, companies have increasingly arranged shift rosters of 60/30 and 90/30. Intra-regional LDC workers usually have shorter rotations such as 7/7 or 14/14. However, they also can be required to work the longer rosters outlined above. The conditions of this kind of multi-locality are set by the different shift rosters. Longer shift rotations are more challenging, as they mean being away from home and being "locked" in the camp for a substantial time. Shorter rotations usually put less pressure on workers. Workers prefer longer off-duty times, however, in the context of longer shift-rosters. The specific shift roster offered to workers depends on requirements at the workplace, job rank, distance and travel time, modes and cost of transport, and the general transport infrastructure in a region. It also depends on company policies to operate cost-effectively, i.e., fewer crew changes mean fewer interruptions in the workflow and lower transport costs.

Smaller companies rarely foot the bill for flying their workers to the site or the nearest hub town. However, large corporations often cover the flights to and from work; Gazprom, for example, runs its own airline (*Gazprom Avia*). Some companies may decide to pay for train travel instead of flights. Many workers reported having to pay for transport to the hub-town out of their own pocket. These workers tend to travel by train, as it is substantially cheaper. The travel time difference between the train and flying may be several days. For example, it takes three-and-a-half days one-way between Moscow and the hub-town Novy Urengoy and only four hours by airplane. Long train journeys thus substantially reduce the recreation period. From the hub-town workers are picked up by company transport and transferred to the remote worksites.

Long-distance commuting is used around the world to provide a skilled labor force in remote areas where the local workforce is not sufficient or properly qualified. Furthermore, it is used for temporary and mobile projects – such as in the offshore oil sector – where no industrial town or settlement is built near the extraction site (Hobart 1979; Krivoy 1989; Spies 2009; Storey 2001). Since the late 1970s LDC has been common throughout the circumpolar North, not just in the Soviet Union. Other sectors besides oil and gas that rely on the LDC system include forestry, fisheries, and mining. LDC in Russia increased from the late 1980s onward (Aleshkevich 2010; Eilmsteiner-Saxinger and Aleshkevich 2008).

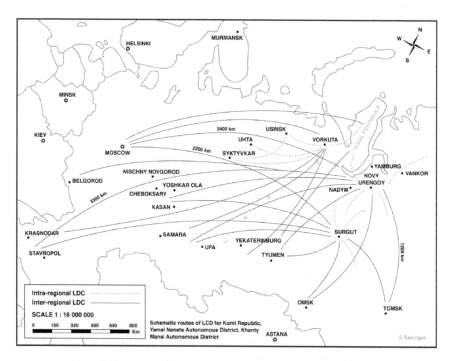

Map 5.1 Schematic routes of LDC for the Komi Republic, Yamalo-Nenets, and Khanty-Mansi autonomous districts.

This mode of labor-force organization allows companies to recruit from a large pool of workers, who can be flexibly transferred any time to any place, as needed. At the same time, LDC offers workers a large job market that is not confined to the home region. This provides substantial job opportunities. The Russian oil and gas sector is highly dependent on a mobile and flexible workforce since the market and the extraction of resources, as well as the exploration of new fields, is highly dynamic with new sites frequently opening and closing. LDC bridges the current shortage of highly qualified personnel in the local labor market through the availability of an interregional pool of workers (cf. Ananenkov et al. 2005; Andreyev et al. 2007; Andreyev et al. 2009; Krivoy 1989).

Employment conditions

The Russian petroleum sector employs around one million people today (Martynov and Moskalenko 2008). The exact number of LDC workers is not known since this category is not covered in the national census and no central registry is available. It is estimated that a few hundred thousand men and women are on the move for employment in the oil and gas industry. For instance, Regnum news agency reported on March 3, 2011, that in the Republic of Bashkortostan alone over 100,000 people are LDC workers. Indeed, around 90 percent of the employees of Gazprom subsidiary Gazprom Dobycha Yamburg are LDC workers (Andreyev et al. 2007). These statistics confirm the high demand for a mobile workforce in this industry, especially during booms and high oil prices.

The production and the construction segments of the oil and gas industry have different employment conditions. While production is stationary and offers fairly long-term work during a production period, the latter is short-term and variable according to the location of the very construction site – be it construction of plants, pipelines, electricity, roads, or other infrastructure. Both sectors, however, can operate with long-term employees, but only the production sector actually does.

Today, the construction sector operates overwhelmingly with short-term contracts. They offer contracts for the duration of a project, that is, only for a few shifts or one season. Long-term job security is therefore rare. This is not specific to Russia, but applies to other regions worldwide. In the Russian case, however, workers who know the job from the Soviet era or entered this sector during the 1990s and early 2000s face a huge change in this respect. Previously long-term contracts, especially with the state and quasi-state enterprises during and shortly after the Soviet period, were the rule. In the reform period of the late 1990s and 2000s contracts have been increasingly divided into smaller outsourced activities. Today, especially in the construction sector, outsourcing to subcontracting and even sub-subcontracting companies is rampant. These private corporations and small businesses compete on the market under neoliberal conditions. A clear pattern has emerged of large companies selling parts of their operations and subsidiaries and laying off employees, who are re-contracted by the new owners under new, less preferential conditions: shorter contracts, lower payment, no or less coverage of travel costs, for example. Many of my interlocutors who had worked for decades under secure job conditions thus feel betrayed – both by the companies and by the state, which they expected to continue to take care of workers. The younger workers accept this as the reality of today. Complaints to management yield the same response: "If you are not satisfied, someone else is queuing up to take the job" (Saxinger 2016a).

The existing legal regulations in this industry cover supplementary payments for shift work in the Russian North, maximum working hours, breaks, and shift length, among other terms. LDC is a separate category in the labor code and differs from, for example, seasonal work or business trips. The law defines LDC as a long-term employment contract based around certain shift rotations of a work period followed by a rest period (Martynov 2010). However, especially since the global economic crisis in 2008, many companies try to employ their workers not as LDC but under other classifications that avoid the legal requirement to pay supplements for shift work. Such employment strategies are a gray area in the law.

Employees in large enterprises, primarily in the production sector but also some large construction corporations, usually enjoy not only long-term LDC contracts, but also higher salaries that are negotiated with the labor unions. Other special benefits include subsidized holidays abroad or typically on the Black Sea, company pensions, stipend programs for the employee's children, social support in case of injury or death, and subsidized loans for housing. Competition to get into such enterprises is therefore high. These fringe benefits are either part of corporate social responsibility or negotiated between the trade union and the company. Many of these benefits are a Soviet legacy: in the Russian North, such as in KMAO and in YNAO, industry restructuring went at a much slower pace than in central Russia (Moe and Kryukov 2010). As such, people still expect special benefits from "their" company. This is especially true for local (commuting and non-commuting) employees from Northern mono-industrial towns, while LDC workers from the central and southern parts are eager to get the substantially higher salaries in the North compared to the average in their home regions.

Salaries in the oil and gas sector in the Russian North are among the highest for this sector. This is in part due to so-called Northern benefits (*severnaya nadbavka*), which are legally defined and represent up to 80 percent additional salary based on the geographical latitude of the workplace (Kozlynskaya 2009). Salaries within each job category depend on work experience, education, trade certificates, and the like. Workers from central and southern Russian regions report that this is in many cases triple or often much more of what they could earn at home. But compared with international salaries in this field, the Russian income in this sector is modest; the highest wages per month that I came across were for camp service personnel ($1,200); heavy equipment operators, truck drivers, etc. ($1,800); and skilled trades ($2,500), with highly qualified engineers and management personnel earning beyond that. In Australia and Canada these resemble approximately weekly salaries.

The lower number of women in the oil and gas sector reflects the discrimination when it comes to employment in this field. Today women primarily are concentrated in lower-paid jobs. Unlike in the Soviet era, where female employment in technical fields was common, today companies are much more likely to hire a man instead of a woman. Young women in technical spheres especially struggle to get a job in such male-dominated arenas. Women are employed instead en masse in administration, the service sector, or in chemical laboratories – typical female jobs – that are at the lower end of the pay scale. Masha,[2] a student at the Ufa State Petroleum Technical University, states:

> Companies from all over Russia recruit directly at the university and the demand for highly qualified engineers is high. But they prefer the guys. You can say that one applicant has the choice of several employers. Even if we have nearly full placement of graduates on the labor market, including women, this does not mean that the payment is high. Especially women are among the lowest paid ranks. However, we are happy to get a job at all, usually in the administration. A female engineer has fewer chances to be recruited, since the company does not want to train somebody who it probably cannot retain, e.g., when she becomes pregnant.

A widely reported phenomenon is "wild commuting" (*dikaia vakhta*). This refers to workers who on their own initiative strike out for the "blue and black (gas and oil) gold rush" in the North without a pre-arranged job. In many cases they are not recruited via the approved agencies with whom the large companies cooperate. They go to the North to find a job on their own or through one of the many Internet-based agencies. Often such online agencies trick clients who usually pay high fees for services that may turn out to be useless. Furthermore, "wild commuting" often puts workers at risk of precarious employment conditions: for example, illegal employment, contracts that do not fulfill legal requirements, and unreasonable workloads that cannot be fulfilled and subsequently are not paid. Cases are reported where the worker must surrender his labor book (*trudovaia kniga*) or passport to the employer or the crew leader and the document is only returned upon satisfactory completion of the job. This traps workers to sites; if any problems arise, they cannot quit or escape from their job, which is located remotely with rare pickups. As a result, interlocutors sometimes speak of "slave-like" conditions. Vadim, a pipeline welder from Bashkortostan, recalls being trapped on a remote site in the Tundra:

> The specified amount of pipeline pieces we had to weld per day was much too high. Not even the most experienced group among us could meet these requirements. Therefore, we got less payment. This was substantially lower than I had expected and what I needed to pay off my bills. [...] Anyway, you do your job, if not you do not get any money at all and have wasted your time. But I was not familiar with these conditions beforehand.

Other cases have been reported where workers have been made to sign a resignation letter at the same time they sign their contract, which makes it easier for the employer to lay off the person. Such illegal mechanisms ensure that no complaints are lodged when salaries are not paid in full or not paid at all. When I asked if they had complained to the labor union or gone to court, laughter was the reply. "Once you complain, your reputation in the sector is down and no one else would dare to hire a troublemaker. Furthermore, you have no chance to succeed in a court case," one informant explained, pointing to the weak judicial system and corruption, as well as to the weakness of the labor unions.

Gazprom Dobycha Yamburg (GDY) was one of the first companies to introduce LDC work in the early 1980s instead of building another mono-industrial town (which was the common form of labor force provision up to that point) (Aleshkevich 2010; Eilmsteiner-Saxinger and Aleshkevich 2008; Pashin 2004). As pioneers in this field, they set up a research department for LDC work and recently launched a "LDC management system," which should improve and regulate not only employment conditions, but also standards at the workplace and in the camps (Andreyev et al. 2009). According to the vice CEO of GDY, Valentin Kramar, the company has lobbied government officials to include key elements of this management system into the labor code.[3] The authors of the "LDC management system" (Andreyev et al. 2009), who have published several related scientific articles and who are employed in or closely related to the company, are critical of the current situation regarding the insufficient legal regulations surrounding LDC that date back to the Soviet era[4] and about the lack of enforcement of existing laws. In an interview with Vladimir Borovikov, who was previously responsible for human resources at GDY and is also a scholar in this field, he states:

> Today this sector operates under the pressure of a globalized petroleum industry. It is not surprising that the working conditions are to a certain extent weak, since the old regulations are not effective anymore and new sufficient ones are not yet in place. Before we can talk about obeying the rules of the game we first have to set and elaborate new rules.[5]

A main concern of LDC workers, especially blue-collar laborers, is the unpredictability of their career. As stated above, sometimes contracts last only a few shifts. While at home from a shift, many workers constantly search for new job opportunities. The lack of job security creates substantial distress and undermines the social security of the individual and their family. Since wages are still attractive in the North, people try their luck again and again, particularly from regions where economic development still lags behind (Öfner 2014; Saxinger 2015; Saxinger et al. 2016). Continuous outsourcing in the oil and gas sector gives rise to unfair players in this field. The combination of a global neoliberal practice, corruption, and a weak judicial system, such as in Russia, creates a structurally precarious situation. The system of "wild commuting" gives the whole sector a negative reputation. As a result, qualified people who can find jobs in other sectors with equal pay are likely to refrain from entering the oil and gas business that requires LDC. The overall qualification level in the industry might therefore plummet as a result, as Bykov (2011) discusses.

Life in the camps

LDC work requires living in two geographical and socially meaningful places: at home and on duty. Most shift rosters dictate that people spend in total half a year or more in the camp. In contrast to "nine-to-five" jobs, LDC workers are under company surveillance 24/7. After a 12-hour shift at the workplace, what passes as private life takes place in the company camp nearby. Full commitment to the camp regulations is part of the contract and usually companies do not hesitate to fire a worker for violation of the rules. Life and work satisfaction are substantially dependent on the conditions in the camps. These relate to the overall construction and architecture of the camp, the layout of the rooms, food supply and fresh water quality, heating, communication infrastructure, social facilities for recreation, and more.

I came across two extremes in terms of camp quality and many variations in between. First, I visited two camps of state companies (one operated by GDY, which I describe in detail below, while the other shall remain anonymous). As extraction and production companies, they run stationary camps, which have very good conditions. Second, I went several times to mobile camps with more tenuous conditions compared to the stationary camps. Among these mobile camps, conditions range from weak to moderate.

Mobile camps are set up to carry out pipeline maintenance, road construction, and other sitework involved in infrastructure construction. Such work requires the gradual mobility of the workplace and so that of the camp. Usually these are smaller camps for around 20 to 50 people, while stationary camps can house several hundred to thousands of people. The examples that I share here are also based on accounts from workers who I talked with and do not refer to my observations alone.

The GDY camp at the Yamburg Gas Field on the Tazovksy Peninsula in YNAO is certainly at the high end of the scale. The camp consists of several thousand people. Accommodation options range from single rooms to dorms for up to four people (as far as I was shown). The rooms are well equipped with Internet access, a TV, and a small kitchen. In the other camp I visited, on holidays employees are invited to decorate public spaces themselves, usually with self-made items and handicrafts, which are often made during shifts as recreational activities. Several gardeners are employed in the Yamburg camp to maintain the numerous outdoor and indoor winter gardens. The complex is light and green with large hallways, which is important, especially during the dark winter months. Facilities include halls for various sports ranging from boxing to ball games, several gyms, and a large swimming pool. Tournaments take place on occasion. Most recreation facilities are open day and night to accommodate both day and night shifts. The library is large and has magazines for men and women and a huge variety of classic and contemporary literature. It also lends out technical materials to those who are pursuing further vocational training. "Up to 50 people use the facility per day and over the weekend it can be more who take out books or read here in a cozy corner," says the librarian. The culture hall is modernly equipped and used for concerts and company events. Sometimes music stars are flown into the camp for concerts. In addition to the canteen, there is a restaurant and a bar where people can gather and celebrate birthdays or just hang out and eat dinner. From time to time, the bar offers live music with singers and musicians who are regular employees at GDY. There is a music club run by a music teacher, where workers can play various instruments together, receive lessons, and have practice space. Several shops in the camp sell food, all types of alcohol, souvenirs and gifts, as well as home accessories and everyday items like shampoo and perfume. Vera, an administrative worker, says that she goes shopping once in a while for entertainment: "It is nice to buy a teacup or a nice bowl for my little camp-apartment." A large Orthodox church holds regular services, and the museum presents the history of the camp and the opening of the Yamburg Gas Field. The GDY public relations department runs a TV station and a newspaper, which provide stories and important corporate information to employees. In my transnational research in this field spanning Canada and Australia, I did not come across a comparable camp that provided so many facilities to staff.

The picture was quite different in one of the small mobile camps that I encountered. Workers lived in worn-out trailers dating back to the 1980s. A little wood stove and an electric radiator – on which people also dry their wet clothes – heat the cabins that house eight people. Bedsheets are not provided by the company. The best workers can hope for in such small mobile camps is a TV room and a strong Internet connection for distraction after their shift. Sanitary facilities are an outhouse and a sauna. People frequently have to cook for themselves in these very small camps. However, this description is of the low end. Although many of the smaller subcontractors buy old housing equipment, the majority of mobile camps today are equipped with up-to-date trailers and offer a canteen. Nevertheless, sanitary facilities are in another trailer; a visit requires walking under the open sky to get there even if it is below −45° C. Modern mobile facilities prevail particularly in larger companies and at newly opened sites. Small mobile camps do not have their own source of drinking water. They need to be serviced with fresh water, which interviews revealed as a major concern. The same can be said about fresh food. In some cases, I came across complaints about the unreliable delivery of goods, e.g., in bad weather conditions the helicopter or the truck might not make it through to the camp or simply the company's ordering of provisions is chaotic. A wife of a LDC worker who faced such problems narrated: "He lost 20 kilograms during his shift. He looked like he'd come back from Buchenwald [the German Nazi concentration camp]." Her husband explained to me: "You just cannot leave. If you are at a remote site, you simply cannot escape. There are no roads." While these conditions are an exception, they do still occur.

Besides making use of onsite recreation facilities – in the camps where they are provided – employees like to go for walks in nature. It is usually the more experienced workers who venture outside the camp in their leisure time, as they know the surrounding region since they have worked at the same site for years. The opportunity for walks outside of the camp depends on company policy and security regulations. "You can get lost in the Tundra. It all looks the same when you look around. Never lose sight of the camp and be aware of bears," advises Kolya, a construction worker. People go berry picking or gather mushrooms. Men like to make jam and preserve fruit or dry mushrooms as recreational activities. This is not a favorite pastime among female workers. Masha, a camp service worker, tells me that she is happy that she does not have any household stuff to do when she is away from home. From time to time indigenous visitors come to the camp and sell fresh fish or reindeer meat. This is highly appreciated by the workers who also take such food back home.

Social life can be difficult. While people who work in the same camp year after year feel like a family, those who hop from camp to camp and worksite to worksite do not experience the same close relationships with their colleagues. Basic courtesy is the key to having a good life in a camp: "You have to take care of each other and respect and deal with the many different characters who you meet at the site," says Masha. "Especially when people do not get on, you have to let each do their own thing since there is no way out." Mutual emotional support is essential, particularly near the end of a month-long (or longer) shift. Tensions can run high by that point since workers are physically and mentally exhausted. Often the more experienced employees informally mentor new workers who are not yet accustomed to this way of life. "It is a matter of getting used to it and trying to take it as your normal life. We live that way and you have to cope with it," is a very common statement. Socialization to this way of life is easier for those workers who come from LDC families or LDC communities where they learn in advance about the living conditions in campsites.

In many cases relatives and friends apply to the same employer and request to work in the same sites. People call this "LDC dynasties." "Companies," as reported by one crew leader, "like to recruit through recommendations since loyalty to the employer is then higher and no one wants to spoil the reputation of the relative or friend who brought you into the company." I came across similar statements among managers. Besides socialization to this lifestyle and the presence of close relations in the camp, workers must be able to have alone time as well. People withdraw to their rooms when they crave some solitude, which is more problematic in the shared dorms. Masha says: "You are the whole day and a whole month with your colleagues. Therefore, you need your privacy. It is good not to get too close and not to get engaged too much in other people's business. This helps you to survive." The need to be alone, which does not mean to be isolated or lonely, is crucial for individual well-being.

Camps are either wet or dry, i.e., alcohol is allowed or prohibited, depending on how companies perceive the effects of alcohol. Wet camps allow moderate drinking (getting drunk is banned) as alcohol may reduce stress levels and it avoids the need to check bags and search rooms for smuggled-in alcohol. Those who run dry camps see alcohol as negatively affecting discipline in the camp and a factor in conflict among workers. At the workplace alcohol is strictly forbidden in any case. People are randomly checked for alcohol and drug abuse. Andrey, a crane operator, complained about the medical exams that are usually required before starting the job and general health monitoring during employment: "We are checked like cosmonauts. This is totally different than in previous [Soviet] times." However, screening for alcohol and drugs as well as health checks are standard worldwide in the industry. That said, in smaller camps this is not necessarily the rule. Drugs are a problem, although they are strictly prohibited, and violation of this rule can lead to immediate dismissal. Nevertheless, after the shift drugs like cocaine are widely consumed. Psychological distress is high among camp residents, and some people suffer from depression. Suicides are also reported, but this tragedy is not exclusive to Russia; it is internationally a problem in LDC shift work but one not too openly talked about and not yet studied in detail.

There are no existing standards governing the social conditions in the camps, either at a local or international level such as by the International Organisation for Standardization (ISO). The existing standards cover primarily construction, fire safety, hygiene, and food safety. Less attention has been paid to the mental well-being of workers, and relevant camp standards would help create meaning-ful social spaces in the camps, an approach that Gazprom Dobycha Yamburg pursues. Regardless of standards, some people quit after a few shifts. Others are on the move all their working life. Good social and other conditions, however, are key if people are to embrace this way of life. Therefore, job retention very much depends on the facilities provided by the company. A sector that struggles from high turnover (Beach et al. 2003) must take this aspect seriously.

Occupational health and safety

Safety is an overwhelmingly important topic for most of the people I met. The bottom line of the discussions refers again to the great diversity of conditions among companies, as well as within companies and their subsidiaries. In general, the large companies and corporations have a sophisticated health and safety man-agement system that is enforced top-down and regarded as sufficient among my interview partners. This is also a matter of public relations since large companies try to avoid having a bad reputation. Difficulties especially arise with the smaller and more remote units away from a base camp. It also depends on whether the work requires handling dangerous goods or operating dangerous machinery or if the work is carried out under hazardous weather conditions.

Workers tend to experience greater problems with smaller subcontracting companies. The top-down enforcement of legal safety requirements is lax and the controls are not regarded as sufficient. Workers appeared to me to be very concerned about safety in the workplace, as they regard their health and physical condition as their most important asset for making a livelihood for their family; social benefits in Russia in cases of injuries and disability are not sufficient to sustain a household. Some companies voluntarily pay settlements in such cases. For the rest, payment is governed by industry-wide collective contracts worked out by the labor unions. Despite not being able to live off disability benefits and injury-related subsidies in Russia, individual workers are unlikely to sue their company in cases of employer negligence. The general assumption among the people I met is that law and labor rights exist solely on paper in Russia.

Insistence on workplace safety, for example ensuring that safety clothing is worn or that safety equipment is used, is not consistently observed in smaller companies. Many workers reported that they are themselves responsible for buying compulsory safety clothing and safety shoes, as these items are not provided by the company. The cost of safety is in this way foisted onto workers. On the other hand, neglect of safety issues reflects also a lack of awareness among workers themselves and their foremen, who are in charge of enforcing safety regulations. Safety procedures also include emergency communication channels between a field outpost and the company administration at the center. This is particularly important in smaller work units where sufficient help for severe accidents is not available locally.

Sergey is from Samara, and I first met him while staying a few days in one of the private flats in Novy Urengoy, which are rented out to LDC workers on their stopover to the camp when they arrive by train or airplane in the hub-town. He was waiting for the helicopter transfer to his workplace at a new construction site a few hundred kilometers beyond the Arctic Circle. His brigade paves earth surfaces and lays roads. Sergey operates heavy machinery that requires considerable experience and skills. He is in his mid-twenties and is foremost concerned about accidents and fatal injuries. Just a few years ago he got married, and he has a two-year-old son:

> Who will care for my family when I am handicapped or even dead? I could only be happy if my wife found as soon as possible a new husband who would care for them. However, thinking of this, I myself, and me and my wife together, wonder if these conditions are really worth the money I earn. [...] I am not sure how long I will go on with shift work.

Sergey recalls heavy machinery tipping over or being lost to the marsh. Even if his brigade is equipped with a satellite phone, in the event of a severe accident a first-aid helicopter still takes some time to arrive and whether it can fly at all depends on the weather conditions. Grigory is a truck driver whose *Kamaz* (a widely used truck type in the North) tipped into the marsh. The vehicle was never seen again. "No idea how I managed to escape, but I am here as you see," he says. He reminded me again of the fact that this is the harshness of the North and the peculiarity of the jobs there. Such circumstances are a matter of fact and "if you want to be part of this game and want to earn good money, this is the reality," Grigory admits.

Safety risks are further posed by the conditions on overland roads (*trassa*). This is of crucial relevance in the winter period, when frost, ice, hail, and blizzards produce treacherous driving conditions. Vasily has worked in the North since the early 1990s and knows the dangers involved in traversing the overland roads: "I have witnessed already a lot of severe incidents. I know people who have frozen to death when help came too late after an accident on the overland road." Similar to Grigory, he is accustomed to the extraordinary working and safety conditions in the North. LDC work involves risks, but new and seasoned workers alike balk at companies that are reckless in the face of these risks. Vyacheslav is from

Krasnodar and the driver of a passenger truck (*vakhtovka*). He just started his first
shift in the North and complains about the safety practices in his company:

> My truck regularly breaks down. At the headquarters nobody cares. It is
> insufficiently maintained and I take it on the road again until the next inci-
> dent happens. [...] I will quit after this shift. I already talked to my wife
> in Krasnodar, and also she says the money is not worth the risk to my life.
> It would be good money, but money is not everything.

His account speaks of a reckless firm that disavows standard safety practices in
this sector. Vladimir, also a passenger truck driver, explains to me the safety procedures in his unit:

> First, you have a checklist and before every departure you inspect the vehicle.
> You get your gasoline etc. and after the passengers have embarked you call the
> switchboard at your target via radio and notify your departure and expected
> arrival time. You check back in with the office for weather conditions. If you
> do not arrive at a certain time a rescue unit will come and pick you up.

The skies also carry risks. A main safety issue is air transport, particularly under
bad weather conditions. People are worried. Sometimes brigades have to wait several days due to bad weather conditions before a helicopter can take off. However,
bringing the crew to the site on time is an essential part of the contract for the competing air service companies. For the air company it is a trade-off between safety
and schedule. Workers, therefore, are concerned that airplanes or helicopters are
not de-iced properly, since this takes valuable time. But snow, frost, or ice stuck to
the blades may lead to severe accidents.

Safety issues are in general a structural problem since the independent and
institutionalized controls in place are not sufficient to ensure that companies follow the legally set safety standards. This applies also to hygiene standards and the
provision of food and housing at the workplace and in the camps, particularly in
smaller, more remote ones. I interviewed someone in charge of enforcing safety
and hygiene standards at worksites in a Northern municipality. She explained that
there are several reasons for the weak state controls. First, the tasks within the
control system are divided among various independent state bodies with poor
communication among the departments and teams and no standardized report-
ing system. Second, the remoteness of the worksites, the scattered system of
numerous small subcontracting companies, and the vast territory involved make
it challenging to conduct tight inspection patterns. Third, surprise inspections are,
in practice, not possible because of the restricted access to these vast industrial
zones. Any individuals or vehicles entering a site are directed through a tight
meshwork of control posts. Notification at the receiving site and registration upon
entrance to the restricted industrial zones is standard. This is, on the one hand, a
positive example where security standards are fully fledged and strictly adhered
to (i.e., the vigorous controls and checks of people and machinery entering sites

hinder potential sabotage and terror acts), but it makes any ad hoc control of the sites nearly impossible.

In general, safety is a high priority but nevertheless there are shortcomings in terms of awareness and implementation of standards. Safety, quality of life, and health are the most important aspects that my interview partners raised when it comes to satisfaction with a workplace and with an employer. Cases show that lack of safety is a major reason for leaving a job. As in the case of the dodgy employment conditions described above, it is a matter of control and enforcement of existing regulations and standards. Furthermore, corruption makes the sufficient control of practices weak.

Conclusion

The petroleum industry and its related sectors in the Russian North, especially in the Yamal-Nenets autonomous district and the Khanty-Mansi autonomous district, provide high salaries in comparison to other economic sectors and regions in the country. However, jobs in this field place extreme demands on workers, considering the harsh climate, remote workplaces, and the itinerant lifestyle divided between home and long periods onsite. Nevertheless, it is not a mobile and multi-local lifestyle per se that accounts for difficulties in retaining the workforce. It rather depends on the conditions under which mobility and multi-locality are implemented. Satisfactory camp standards and conditions are the key to successful LDC work. Since people spend "half of their life" onsite, this space needs to be invested in to make it their social and emotional place. Different rules, different social settings, all-day surveillance, and different ways of behavior than at home require a high level of adaptation. However, people turn this way of life into their normality (Saxinger 2016a, 2016b). The extreme work and life in the camp are turned into routine and everyday practice. This is of utmost importance when it comes to a long-term LDC career. The capacity for individuals and their families to cope with this multi-locality depends also on the conditions in camps.

Employees in large corporations in the production and construction sector tend to feel comfortable in their mobile jobs. LDC becomes the norm, as many of my interlocutors put it. After many years in this field, dropping out is no longer a question. Besides already being used to the mobility, separation from home, and living in camps, their work is sweetened by several incentives such as an income above the average, company pensions, and other social benefits. Furthermore, children and spouses are already attuned to this rhythm of work and high household budget. Being permanently on the move becomes an essential part of the individual's life and of their personality. Many cannot imagine returning to a sedentary lifestyle and working "nine to five." Some told me that they "get nervous when being too long in one place," that they need to feel that "the suitcase is packed." This is especially hard for mobile people who are nearing retirement age; many of them continue to work beyond retirement as long as their health is adequate. People negotiate and integrate the social spaces of being at home and onsite (Saxinger 2016b).

They create the camp as a place that they belong to (Eilmsteiner-Saxinger 2013). An erratic lifestyle – as it is perceived from the outside – turns into a stable life through routine and everyday life at familiar places with familiar, but different, social settings. This approach prevails primarily among workers who are employed in state companies or large corporations and their subsidiary companies. Such companies design their camps with an eye to the social and psychological well-being of staff (e.g., light, green complexes with ample recreational facilities) and ensure high safety standards in order to maintain the labor capacity of their human resources. For these workers many of the downsides of long-distance commute work are compensated for by the good labor and living conditions. A different picture frequently occurs when it comes to "wild commuting" or employment in one of the smaller subcontracting companies. Many people hired in this way are not flown in for work at the company's expense, but have to accept long train journeys that cut into their scheduled home leave.

Comprehensive standards for camp accommodations and recreational facilities do not exist. Moreover, in terms of labor law LDC is insufficiently regulated (Andreyev et al. 2009). Unsatisfactory labor and living conditions, as described above, tend to discourage people from making sustainable and long-term careers in the oil and gas sector, which increasingly relies on LDC work. This is a structural problem. Subcontracting companies in the highly scattered construction sector have a problematic reputation in terms of lax control and poorly enforced labor laws. This sector, in contrast to the production sector, still requires huge numbers of blue-collar workers. The results of my study show that workers from socioeconomically disadvantaged regions in central and southern Russia are more likely to become trapped between accepting harsh conditions or ending up less paid or without a job at home (Öfner 2014, Saxinger et al. 2016).

Currently, the sector faces high levels of outsourcing and subcontracting under neoliberal conditions and needs to address the extra risks for workers that this might bring. On the one hand, the Russian state is increasingly taking control of the country's resources, but at the same time it has declined to exercise control over working conditions across the sector. It is not sufficient to see this development as simply "post-socialist" and a product of the early days of the Russian Federation. The Russian petroleum industry describes it as "re-socialist neoliberalism." Under this system corruption is unchecked by a weak judicial system, a combination that leads to high dissatisfaction among employees. This contributes to the high labor turnover in a sector that is dependent upon qualified people who are willing to take up a life on the move. Many people get used to the mobile and multi-local lifestyle and call it their "normality under extreme conditions." Unsafe labor practices and poor living conditions, however, lead people to leave LDC work after a few shifts or a few years. The industry offers prospects of social mobility to some, but at the moment workers, especially subcontractors, are avoiding mobility because of the many potential risks related to this lifestyle. Large corporations with high labor, living, and safety standards, in turn, appear to be long-term, sustainable employers.

Notes

1 This study is funded by the Austrian Science Foundation (FWF) [P 22066-G17] in the framework of the research project "Lives on the Move" (project leader: Heinz Fassmann, 2010–2015) at the Department for Geography and Regional Research, University of Vienna and at the Institute for Urban and Regional Research (ISR) at the Austrian Academy of Sciences: raumforschung.univie.ac.at/forschungsprojekte/lives-on-the-move. Funding was also provided by the University of Vienna, Austrian Research Association (OEFG) and the Austrian Academy of Sciences (OEAW). Furthermore, this publication is related to the project "CoRe – Configurations of Remoteness," funded by the Austrian Science Foundation (FWF) [P 27625-G22] (project leader: Peter Schweitzer, 2015–2018). This research is also related to the ESF/Academy of Finland funded project BOEAS-MOVE INNOCOM at the Arctic Center in Rovaniemi, Finland (project leader: Florian Stammler, 2006–2010). An earlier version of this chapter was presented and discussed at the Arctic Frontiers Conference in Tromso, Norway, in 2014, within the panel on Arctic migration organized by Marlene Laruelle (GWU, Washington, DC) and Aileen Espiritu (UiT, Tromso) and funded by the Norwegian Research Council. Special thanks to the numerous informants and experts in Russia as well as to Elena Aleshkevich, Heinz Fassmann, Elena Nuykina, Elisabeth Oefner, Florian Stammler, and Peter Schweitzer for discussion, to Jenny Rasell for proofreading, and to *OOO Gazprom Dobycha Yamburg* for providing access to the Yamburg commuter camp and their archives.
2 All names are pseudonyms, unless otherwise specified as interviews with experts.
3 Author and Elena Aleshkevich, interview with Valentin Kramar, vice CEO of Gazprom Dobycha Yamburg, responsible for human resources and social affairs, Novy Urengoy, 2010 (Project "Lives on the Move").
4 For an overview of the current laws on LDC, see Martynov 2010.
5 Elena Nuykina, interview with Vladimir Borovikov, former head of the human resources division at Gazprom Dobycha Yamburg, St. Petersburg, 2012 (Project "Lives on the Move").

6 The urbanization of indigenous peoples of northeastern Siberia

Vera Kuklina and Natalia Krasnoshtanova

This chapter examines the process of urbanizing the indigenous peoples of the Russian North in different types of settlements. Unlike urbanization in other countries, the Russian case features a more preserved traditional culture (mostly nomadic and shamanistic), a strong Soviet heritage in the man-made environments and educational system, and harsh climate conditions. The highly centralized spatial organization of the USSR (and, later, of the Russian Federation) has influenced the locations of indigenous activists, who are mostly based in Moscow. Here, it is crucial to see if there is any leverage for indigenous peoples to voice their own concerns and to be heard or to make changes at the regional and local levels, especially concerning large industrial projects. In order to understand this issue, we present case studies based on fieldwork and secondary sources in the Republic of Tyva, the Republic of Sakha (Yakutiya), and the Baikal region. Specifically, we will discuss migration to the cities, urbanization of rural areas and traditional use of nature, and the impact of the extractive industry in terms of transforming the traditional lifestyle into an urban one, including development of urban infrastructure, employment patterns, and communication between the village and the city.

Although over half of the indigenous population lives in cities, there have been few studies of the problems associated with urbanization (UN Habitat 2009). Such urbanization requires not only adaptation to a different way of life, but also adaptation to a different cultural environment and, sometimes, adaptation to another language (Lobo and Peters 2001; Kishigami and Lee 2008; Lobo 2001; D'Arcus 2003). This problem is especially relevant for the Northern and Arctic regions, which have the highest proportion of indigenous peoples. However, the question of what constitutes this type of urbanization in terms of changes in the geographical space is little explored.

The analysis presented here utilizes interviews and field observations in the Katanga region and in the municipality of Tofalar Nizhneudinsk district, Irkutsk oblast, Todzhinsk district (Kozhuune) of the Republic of Tyva, and expanded interviews from Bauntov district of the Republic of Buryatiya, Trans-Baikal krai, Republic of Sakha (Yakutiya), and the cities of Irkutsk and Ust-Kut, as well as statistical data, censuses, and official documents collected by the authors in 2006, 2008, 2013, and 2014.

The chapter examines changes that have occurred in recent years in the urbanization of indigenous peoples. It looks at the changes in the way of life, employment, and communications under the influence of infrastructure transformation and the renewal of the mining industry. It explores how urban indigenous peoples and old Northern communities are involved in the problems that their countrymen face due to the development of the mining industry. Given the paucity of indigenous and "historically indigenous" peoples, as well as the inability to highlight "historically indigenous" peoples as the statistical group in the study, we operate with data from interviews, which we compare with the statistical and other official data at the municipal level. Sometimes we deal with individual cases that show a variety of strategies and opportunities associated with urbanization of the indigenous peoples.

Changes in the urban population

Statistical data change depending on the manner of defining the indigenous peoples. In Siberia, the concept of indigenous peoples is ambiguous: officially recognized as indigenous peoples in the international documents are representatives of the indigenous peoples of the North, Siberia, and the Far East. This list includes both the "historically indigenous" (Bogaturov, Dundich, and Korgun 2011) peoples of Siberia (Yakuts, Buryats, natives of Khakassia, Altai, and Tyva), as well as the "historically rooted" (Rozhansky 2002; Turov 2008) peoples; that is, the descendants of the Cossacks and peasants who arrived in Siberia a few generations ago and had time to "take root" in the local environment. This is especially true for the south of Siberia, where in the seventeenth century the indigenous peoples and migrants from the western regions of the country were mutually borrowing technologies, economic strategies, and techniques of existence to survive in the harsh and fragile environment (Stammler 2010).

If we take into account only the group of "minority indigenous peoples," their population is 316,000 people, or 0.22 percent of the total population of Russia. When taking into account the "historically rooted" peoples, the total number rises to 1.67 million, but it is still just over 1 percent of the total population (2010 Census). Between the censuses of 2002 and 2010, urbanization increased among the "historically indigenous" peoples of Siberia from 39.7 percent to 43.5 percent, and among the minority "indigenous peoples" from 32.2 percent to 33.7 percent, while the total growth for Russia was from 73.3 percent to 73.7 percent. Including the "historically rooted" peoples is challenging, as it requires the study of biographies of local residents, but it should be noted that, according to research conducted by Timothy Heleniak, a sense of attachment to a place is also formed among the first generation of migrants who have lived here for a long time (Heleniak 2009b). In addition, the growth of international migration may strengthen the concept of "indigenous peoples" and define them as all locals who are in opposition to migrants (Laruelle 2014).

One of the long-standing disputes among the urbanists is formed around the definition of "the city." The statistics in Russia consider settlements to be urban, if they are approved by legislative acts as cities and towns (working towns, resort towns, and holiday-cabin villages).[1] From an ethnographic point of view, the city, in the Russian tradition, is a "local economic and cultural center, a relatively large settlement with a more complex social and ethnic composition of the population (than that of rural settlements), most of whom are employed in the production for exchange and in the exchange, which generates a set of features of home and social life, which is a feature of the urban lifestyle" (Rabinovich 1983, 24). The urban lifestyle emphasized here can be found by sociologists in the post-socialist village that has experienced industrialization and modernization (Oswald 2012). On the other hand, studying the "household" urbanization (heating, hot water, sewer systems, waste management, and electricity) and the density of the population, demographers have come to the conclusion that there is an insufficient level of urbanization in the existing urban areas (Karachurina 2012).

The definition of the city varies from country to country; thus, we often take into account measurable indicators, such as population. According to research by Lee Huskey (2010), small settlements (less than 1,000 people) make up over 83 percent of the settlements of Alaska, the Northwest Territories, the Yukon, Greenland, and Iceland, but are home to only 15 percent of the population of these regions. Based on the estimates of Larisa Riabova (2010), in Russia indigenous peoples usually live in small settlements, but the country displays the highest concentration of population in major cities in the Far North and equivalent areas – more than 80 percent. However, the general trend of concentration in larger settlements is typical for them as well (Heleniak 2009a).

According to the observations of Susanne Dybbroe (2008), there is an increased level of urbanization among the indigenous communities, in the sense of quality of life and the urban experience, while such urbanization features as size, density, and structure of the settlements remain doubtful. In the study of the transformation of indigenous lifestyles, ethnographers usually define three groups: the urban population, rural and nomadic population, and people living outside of permanent settlements (Sirina 2009; Habeck 2002). However, if one follows the conception of urbanization not only as statistical data, but also as a change in the lifestyle and concentration of the population in settlements, this process can be traced in all three groups. Each of the groups, as well as the relationships between them, will be discussed in detail further.

Urbanization in statistical indicators

Unlike other countries in the North, where the urban representatives of indigenous peoples are characterized by low levels of education, educational migration remains the most common method of migration from rural to urban areas in Russia, especially among the indigenous peoples. This is partly due to the Soviet legacy: Schweitzer and Gray (2000) note that as a result of "indigenization" (*korenizatsiya*), political elites, teachers, doctors, and professionals in cultural fields had settled in the villages and towns of Chukotka. Andrey Petrov (2008) notes that the level of higher education among the indigenous peoples of the North is much lower than the national average. These observations are also valid among the peoples of Eastern Siberia. However, according to the 2010 census, the proportion of persons with higher education among the urban indigenous representatives at the age of 15 years or more is at least two times higher than that of the rural population (see Table 6.1). In particular, the share of urban people with higher education among the Evenki is 17.6 percent, while for rural people it is 6.7 percent; in Irkutsk oblast, in the Republic of Sakha, it is 21.5 percent and 10.2 percent, respectively, and in Krasnoyarsk krai, 18.8 percent and 4.5 percent. It is difficult to use statistics to evaluate the Tofalars and Todzhins due to the scarcity of their population: only two Todzhins were recorded as city dwellers, both with higher education; the number of urban Tofalars was 36 people, of whom five have a higher education, while among the remaining 461 Tofalars, 12 people with higher education were registered. Thus, given the very low levels of education, urban indigenous peoples significantly stand out.

In Irkutsk oblast educational migration is primarily directed toward the city of Irkutsk, which has a high concentration of universities, colleges, and other schools. With this in mind, young men studying or working in Irkutsk often come home to hunt for food and profit during the summer and winter holidays for an extended vacation.

During the Soviet era, the Institute of the People of the North in St. Petersburg was the main channel for the formation of local intelligentsia according to the policy of "indigenization" of local elites; that is why most teachers and representatives of local authorities from the indigenous peoples of the North are its alumni. Currently the Institute of the People of the North, the Polar Academy, and the Institute of Technology and Design in St. Petersburg offer courses on the language and culture of the indigenous peoples of the North; there are quotas allocated to representatives of the indigenous peoples. Krasnoyarsk Medical University trains most future doctors among the representatives of indigenous peoples, and it has preserved the Soviet-era quota system.

If the Institute of the People of the North initially focused most of its efforts on providing training in the Russian language for the indigenous elites, now these efforts are redirected to the preservation of indigenous languages, which are gradually disappearing from everyday use. Thus, out of 1,271 Evenki in Irkutsk oblast, only 110 stated that they could speak the Evenki language; out of 678 Tofalars, 81 people stated that they could speak the Tofalar language, and only a few individuals noted that they could not speak Russian. The situation is less dire in the national republics: in the Republic of Tyva, where Tyvan remains the main language of communication, 1,320 out of 1,852 Todzhins stated that they speak Russian, but we can assume that every one of them speaks the Tyvan language as well. In the Republic of Buryatiya, 3,291 Soyots reported that they speak Russian, while 3,301 stated that they speak Buriyat. Thus, along with the processes of Russification there also occur the processes of Tyvinization, Buryatization, or Yakutization. Therefore, migration outside the republic may be more difficult for them than migration within it.

Table 6.1 Population of indigenous peoples of eastern Siberia at the age of 15 years and older, disaggregated by region and level of education

| Regions | Representatives of minority indigenous peoples | Population size of the respective ethnic group | Those who indicated a level of education | Including | | | | | | | | Without any primary education | Did not mention a level of education |
| | | | | With professional education | | | | | With general education | | | | |
				Postgraduate	Graduate degree	Incomplete graduate degree	Secondary education	Primary education	Completed secondary educated	Primary education	Elementary education		
Republic of Buriatia	Soyots	2532	2532	14	539	138	629	26	730	344	85	27	0
	Urban	183	183	3	66	22	38	1	40	11	1	1	0
	Evenki	2162	2161	16	283	108	394	131	682	444	86	17	1
	Urban	690	690	10	142	60	156	30	162	112	16	2	0
Irkutsk oblast	Tofalars (Tofa)	497	495	0	17	7	33	29	115	198	85	11	2
	Urban	36	36	0	5	3	9	4	6	8	1	0	0
	Evenki	995	995	4	101	35	197	39	227	270	103	19	0
	Urban	312	312	2	55	19	71	11	77	60	15	2	0
Republic of Tyva	Tozhu Tuvans	1295	1295	1	75	5	207	129	360	407	92	19	0
	Urban	2	2	0	2	0	0	0	0	0	0	0	0
Transbaikal krai	Evenki	1030	1030	4	78	22	189	28	270	294	130	15	0
	Urban	174	174	3	36	3	41	10	46	22	12	1	0
Republic of Sakha (Yakutiya)	Evenki	14938	14938	115	1997	642	3396	1008	4598	2284	763	135	0
	Urban	4221	4221	76	906	318	1071	141	1118	470	97	24	0
	Dolgans	1281	1281	9	150	54	216	154	356	243	86	13	–
	Urban	209	209	8	47	24	43	8	57	21	1	–	–
	Chukchi	477	477	1	27	12	102	22	139	145	27	2	–
	Urban	198	198	1	18	5	52	9	63	45	4	1	–
	Evens	10850	10850	49	1274	501	2421	623	3629	1747	508	98	–
	Urban	3909	3909	38	754	288	1019	127	1086	448	118	31	–
	Yukagirs	870	870	12	116	48	189	36	260	170	38	1	–
	Urban	389	389	9	83	32	104	6	102	50	3	–	–

Source: 2010 Census.

Barriers to higher education include the lack of teachers in remote areas, as well as the fact that few teachers have professional training in modern information systems or interactive teaching methods. Also, rural students often prefer to drop out of their studies after receiving basic education. Approximately half of the rural population has incomplete primary education or less (67.5 percent of Tofalars, 50 percent of Todzhins, 50.2 percent of Evenki in Irkutsk oblast). This is especially true among students from distant and often ethnic villages, where there are no schools. In order to receive a high school education, the children have to attend boarding schools, where living conditions are quite different from those at home – not all children are prepared for this. This is especially true for the boys who prefer hunting to learning; as a result, women outnumber men among the urban population (see Table 6.2). However, as researchers have pointed out, there are more women than men in Russia in general, especially among the indigenous peoples, due to the high mortality rate among men (Petrov 2008).

There are cases when students drop out of high school or college and return home, unable to cope with the urban pace of life. Sometimes there are other reasons: "There are two girls entered a university and came back pregnant!" a 50-year old woman told us. "They gave birth and live with their parents now, what can you do?"

Table 6.2 Population of the minority indigenous peoples of northeastern Siberia, listed by largely populated areas and disaggregated by gender

Regions	Minority indigenous peoples of the North	Urban and rural population			Urban population			Rural population		
		Men and women	Men	Women	Men and women	Men	Women	Men and women	Men	Women
Republic of Sakha (Yakutiya)	Evenki	21008	11086	9922	5486	2908	2578	15522	8178	7344
	Evens	15071	5579	9492	5077	1550	3527	9994	4029	5965
	Yukagirs	1281	627	654	559	257	302	722	370	352
	Dlgans	1906	850	1056	260	94	166	1646	756	890
	Chukchi	670	337	333	262	136	126	408	201	207
Irkutsk oblast	Tofalars (Tofa)	678	302	376	41	11	30	637	291	346
	Evenki	1272	560	712	347	133	214	925	427	498
Tuva Republic	Tozhu Tivans	1856	880	976	2	2	–	1854	878	976
Republic of Buryatia	Soyots	3579	1757	1822	229	99	130	3350	1658	1692
	Evenki	2974	1438	1536	882	395	487	2092	1043	1049
Trans-Baikal krai	Evenki	1387	668	719	208	93	115	1179	575	604

Source: 2010 Census.

In addition to educational migration, the region has experienced out-migration in the 1990s, caused by the fact that the first-generation migrants returned to places of their previous residence – western Russia, Ukraine, and so on (Heleniak 2009b). For the old-timers in the Katanga region, migration that is not related to education is limited to the neighboring cities: Kirensk, Ust-Kut, Ust-Ilim, Yantal. Connections with these cities are being established in part through professional networks, but more often through relatives. Among the reasons for leaving, people listed search for work, the need for schools and hospitals, and sometimes all of these reasons combined:

> Zhenya is already working there and Volodya is working with her on this, how they call it … on the tube. And Olga had to get out of the village anyways, because the kids are small, yes and Lenka is sick there, she is deaf, carrying her everywhere, in May, again they will go to Irkutsk to take her to the hospital. … Zhenya also says that she wants to come home, home, home. Where will you go? Now Ilyusha [grandson] finished 10 classes, right now he is in the 10th grade now, he, too, needs to learn.[2]

Having obtained higher education, a person is less likely to return to the countryside. Among the reasons for people with higher education not to return, locals cite the lack of housing for young professionals and sometimes a lack of suitable jobs. When we asked our interviewees, "Do your young people leave for the city?" one Evenki woman responded:

> Yes, they do. What conditions do we have for them? If someone comes here, he would be given some sort of shack to live in. And we have some serious weather conditions – freezing below 50 degrees, no water and you do not have any firewood. Living in the North is difficult. If one can survive here, it is only due to his huge desire to live.

As noted by researchers in Greenland and Nunavut and confirmed by researchers in Alaska, there is a tendency among the indigenous peoples to return home at the end of the career or for other reasons (Huskey, Berman, and Hill 2004). Similar cases are observed in Katanga district: "I came back and lived near Irkutsk, worked at the medical and obstetrical station," said the chairman of the community. "Then in the late years of my life I was so drawn to home, that I packed and left within a month." Together with other professionals, who studied in the cities, such residents bring an urban lifestyle to the traditional life of the indigenous peoples.

Urbanization of rural settlements

Rural settlements should be differentiated as more populous and less populous. Even without city status, rural settlements that are central in relation to the rest of the territory become a center of attraction of the indigenous population from smaller settlements. Researchers have noted that in the 1990s, the economic changes in Chukotka led to a greater concentration of the indigenous population in larger settlements, which have higher standards of living, more reliable supply sources, and more jobs (Huskey 2010). Traditionally, Northern communities were small, scattered over a wide area, and mobile in order to cope with the resource conditions. Today it is difficult to maintain this lifestyle, as it increases the value of the market products and cost of provision of social services.

Currently, in rural areas populated by representatives of minority indigenous peoples there is an increased range of non-rural types of employment and urban lifestyles, which can be considered a sign of urbanization. Changes have occurred in clothing styles and housing design, and in the presence of household appliances, communication technologies, and transport.

> Almost everyone has freezers. Now we have become more modern. Now we have automatic washing machines! We already have showers and toilets! More or less!

QUESTION: Do you have private water jacks then? For pumping water?

ANSWER: Basically everyone has their own water jack now. Did you see how few wells there are? The one here is broken. They call them water pumping stations. If earlier there were pumps, just pumps, then electric pumps. Since last year, or the year before last, it has become a fashion to have a water pumping station. There are TVs everywhere. You probably saw the satellite antenna dishes, when you land there are only these dishes above the village.[3]

The further technology develops, the more resentment of limitations, associated with the traditional lifestyle, there is among the indigenous peoples:

> A correspondent arrived, either from Moscow, or from … also the men were sitting here like so in their rubber boots, they are wearing rubber boots in the winter and in the summer. They are very comfortable for walking, and are an acceptable wear. "Are you guys wearing rubber boots over sneakers?" [asked the correspondent]. He also was wearing sneakers. Everyone was taken aback, they couldn't understand him. And they say, "and you're not wearing bast shoes?" They say he was blushing. Well, why should we alienate from each other? If you're comfortable in sneakers, why can I not be comfortable standing in sneakers with you? Why should I be standing in high fur boots in the summer?[4]

One of the main incentives for lifestyle changes is the opportunity to interact with geological expeditions and mining companies, which results in the typical urban differentiation of labor and the growth of social and cultural diversity, leading to an increase in the anonymity of the community. In addition, the social infrastructure (kindergartens, schools, clubs, hospitals) accompanying the transition of indigenous peoples to sedentary lifestyle were often maintained by the exploration expeditions.

In Katanga district geologists initially explored new areas with the support of local residents, hiring indigenous people as guides and using reindeer to deliver goods. Under the influence of expeditions, the first changes occur among the nomadic areas – the Evenki are pushed into more remote areas:

> My parents and I wandered on Chon River. Now there is no one there, before there was the expedition there. Now Zagvozdin goes there.

QUESTION: Is he Russian?

ANSWER: Well, yes, they go there. Previously it was Surikov, his father.[5]

The Preobrazhenskaya oil and gas development plant was based in Erbogachen, where a separate district was established. The Nepsk geophysical expedition led to an urban-type settlement – Nadezhdinsk. The settlement was equipped with the latest technology, including computers, starting as early as the 1980s (Kontysheva 2007). With the departure of expeditions, maintaining the infrastructure left behind became too expensive. The settlement was abandoned and officially closed by 2001. Today, there is nothing left from the settlement: "In Nadezhdinsk the settlement grew to about 3,000 people," according to a former district leader. "With heating, hot water, etc. – it was civilized. In a three-year period it was pilfered, it no longer exists – it was dismantled, taken apart, ruined, some things were burned."

In addition to these two major expeditions, numerous smaller ones were based in the towns and villages of the district from time to time, leaving behind scrap metal and building materials that could be utilized by the villagers. Contact with the expeditions encouraged some locals to seek alternate skills and occupations.

QUESTION: Why did you decide to study to become a tractor driver?

ANSWER: As a child, when I wasn't even four years of age yet, I was given a ride on a tractor, back in my native village of Kazimirovo, and that was it. Then, in the 10th grade, there was a drill here, I used to watch how they would raise a tower and fix it in its place. On a 100 [tractor model] the man yanked, pulled up centimeter by centimeter – this here was a jeweler-type of work, now imagine doing that on the 100 and conducting such precision work with it, I couldn't wait to learn how to do that.[6]

In the 1990s, fuel shortages in the large settlements of Katanga district led to the exploration of the nearby oil fields (Yaraktinskoe, Danilovskoe, and Markovskoe). Licenses were issued to companies that committed to deliver the crude oil and gas at discounted prices to the boiler plants of neighboring villages. Thus, diesel power plants and boiler plants were supplied with partially recycled oil. There were also some attempts to use this oil for snowmobiles and motorboats, sometimes at great risk, as the mixture was explosive. In 2000, the Ust-Kut district government attracted investors and co-created the Irkutsk Oil Company, which soon took control of the UstKutNeftegaz and Danylovo Oil Companies, which had been created in 2007. Successors of those companies – Irkutsk Oil and the joint-stock Dulisma Oil Company – are still perceived as local companies, unlike other firms that entered the local market later.

According to a conversation with the 20-year-old daughter of nomadic Evenki, the impact of drilling was not as extensive:

QUESTION: And when they are starting to drill, do you have to migrate or what? Do they interfere?

ANSWER: Well, in principle, no, they do not interfere. On the contrary, they help, for example sometimes they can give us gasoline.

The Eastern Siberia–Pacific Ocean (ESPO) pipeline, built in 2008, has spawned several large enterprises and an unprecedented level of infrastructure development. Construction and reconstruction of power lines and substations to connect ESPO to power grids gave many Northern settlements the opportunity to replace diesel generators with cheaper energy from hydroelectric plants ("Yakutskiie stantsii" 2013). When ESPO began operations, the number of contracts for geological exploration grew 50 percent. According to local estimates, there are about 15 active employees of geological expeditions in Katanga district alone.

Verhnechonskneftegaz Oil has built an all-season road from the field to the Vitim-Peleduy highway, located along the Lena River. They use the airport in Erbogachen village while setting up the winter road from the field to the village (170 kilometers). Irkutsk Oil has built an all-season road and pipeline connecting the Yarakta oil field with ESPO in Ust-Kut. Food and goods are supplied from Ust-Kut to the field. Also, Verhnechonskneftegaz and Dulisma have built pipelines from their gas fields to the ESPO.

For some villages, plowing winter roads was a major problem, because the residents themselves had to organize the effort. Now the industrial companies provide logistical support, which the locals greatly appreciate.

QUESTION: What are the positive aspects of the current changes, of the development of industry here?

ANSWER: ... of course, some type of support is provided. For five years now, or maybe ... no, for the past five years now they clean the winter road.[7]

Naturally, the quality of such roads, made by the specialized machinery, is much higher.

Understanding the relations between the mining companies and local communities is difficult due to the limited communication and underdeveloped transport infrastructure, but researchers have particularly noted the lack of representation of the indigenous peoples (Fondahl and Sirina 2006; Yakovleva 2011; Metzo 2009). Environmental organizations usually do not go to such remote areas due to the limited budgets and lack of skills for communication with remote populations (Henry 2009).

This oversight, intentional or not, is the most remarkable in the case of the ESPO construction. Its original plan called for a pipeline running along the shore of Lake Baikal, which prompted a wave of protests in the cities of Irkutsk, Ulan-Ude, and others (Titov, Rozhanskii, and Ielokhina 2007). ESPO revised its plan to avoid the lake and the most populated areas, and instead run through Yakutiya and Northern Irkutsk oblast. In the Republic of Sakha (Yakutiya) public hearings were held in the cities, while the rural areas were more exposed to the impact of construction (Yakovleva 2011). In 2009, the ESPO pipeline was laid through the channel of the Lena River – one of the largest tributaries of the Arctic Ocean in Siberia (Save Lena Organizing Committee, www.savelenariver .org). Two oil spills from the pipeline have already been recorded in the Republic of Sakha (Yakutiya), and the possibility of other environmental damage is regarded as very high (Pacific Environment 2013).

Irkutsk Oil and Surgutneftegaz have developed websites and phone hotlines so that ordinary people can ask questions or apply for a job, but local residents rarely use these channels. Up until 2013, telephone service was available only in the villages with one or two satellite payphones. The cost of intra-regional telephone calls was 30 times higher in remote areas than in Irkutsk. Prices dropped in 2012, and now are "only" nine times more expensive. Vintage Soviet radio stations remain the primary means of communication between hunters tracking sable and squirrels.

In 2007, all secondary schools gained access to the Internet; in Erbogachen this was done through telephone connection, as fiber-optic communications are not available in the Northern areas. Theoretically, local residents have free access to the Internet, but in practice they do not use this opportunity because they lack basic computer skills or they prefer to connect through their own phones. Gradually, however, online resources are being adopted, particularly with information support from the Russian Association of Indigenous Peoples of the North (RAIPON) and other non-profit organizations. Local indigenous leaders have become more confident expressing themselves as political actors in the negotiations between the mining companies, government agencies, and local communities.

There is not much interaction between the gold miners and local residents. The Pokrovsky gold mine, located 80 kilometers from the villages of Alygdzher and Upper Gutara (with Tofalar population), has been around since 1780; some 5,000 people worked there during World War II, but the local population was scarcely involved. Today, there is no demand for such a large number of people as heavy machinery has replaced hard labor, and only about 300 non-locals work there.

Geological villages could also exist independent of existing settlements, although they would still need to cooperate with those hunters whose land was in close proximity. About 100 kilometers away from the villages of Alygdzher and Upper Gutara, the Tyopsa geological village was established, with all the necessary infrastructure. Only two preexisting houses remain; these are occupied by local hunters. The Zashihinskoe rare metal depository is located in Tofalaria, 60–65 kilometers away from Alygdzher along horse trails. One geologist told us:

> Here are the lands of hunter Nikolai. ... Our leaders have known him for over 30 years, and we maintain a normal and friendly relationship with him. ... A man goes home from the hunt, hunting ended in March, yes, we give him a car. He loads his load on the vehicle, and we take him to Alygdzher. Twenty years ago when we were working here he helped us out. He gave us horses, worked as a musher here, gave horses, gave rides to geologists, and brought samples from distant sites. Here we have distant sections where the vehicle cannot go through. There are geological samples there, he went there on his horses, loaded these samples, and brought them onto the road.

Here, just as in the other communities with mineral resources (particularly, the Bauntovsky and Oka regions of Buryatiya) development is described in terms of "symbiosis" between local hunters and the mining industry: if geologists and gold miners work mainly in the summer, the hunters work in the winter. Thus, hunters have not had to make any significant lifestyle changes. But, in the Bauntovsky region, where the crew is located near the settlements, locals told us that they involved on a seasonal basis: "In the summer they are involved in gold, and in the fall they go hunting." This is significantly different from the oil and gas sector, which is in production throughout the year and even increases its pace in the winter, using winter roads, a time when hunters are also active.

In Todzhinsk district, the Kyzyl-Tashtyg polymetallic deposits that were explored back in 1974 now became a mining site explored by a Chinese-owned firm, Lunsin. The situation is complicated by the presence of Tozhu Tuvan natives, whose ethnic difference may serve as an additional barrier between them and the Kyzyl facility.

In Ust-Maiskiy district of the Republic of Sakha (Yakutiya), the Solnechnyi village is one of the few that survived to serve the gold mining industry. One retiree explained:

> This is an industrial zone, they only mine for gold there, no infrastructure should be there, and no such structure is provided. Because our population is small, all the structures in the area are located here. Five gold miner villages have closed. Bryndakid – a nearby village – has closed, Egorenok closed, Ilikchan, Ustie, Alahmid have closed; they gave people the certificates and evacuated everyone. And they stopped on us and said – this is it! They said we will not be closed, as there would be no one to protect the area.

QUESTION: Do you know how many people live in the village?

ANSWER: It used to be over 2,000 and now is no more than 700, and this is the number of the officially registered, but not all of them live here. I think they will give out the [residency] certificates. 150 families probably reside here. But we have comfortable apartments. Baths, toilets, we have all of that. We have a good settlement. It was the only well-developed village in Yakutiya. Water intake, treatment plant for water, for sewer, heating systems, heating mains – it's all in place here. This is a village since 1964!

QUESTION: This is interesting, because in the north of Irkutsk oblast there are constant problems with freezing over, then something else …

ANSWER: Well, it all depends on the owner. We has such great experts, all were masters with "golden hands"! But in Ust-Mai, for example, there are even no welders; all are alcoholics, or hucksters. They can't weld, they can't do anything! You can call it a big village, naturally!

QUESTION: And you go there by horses?

ANSWER: No, on snowmobiles, on boats, on water cannon vehicles. We do not have horses, because ours was not a farm-type settlement. Therefore, no one is engaged in this here. We have apartments here. This is an industrial zone. We do not have fodder for horses here. There is nowhere to feed, and graze.

QUESTION: And how do you import petroleum then?

ANSWER: They bring petroleum by cars. There is a road there, but every year it gets washed off. And before it never got washed off, and now it does. It is not a highway, more of a primed road. In the Soviet period it used to be spiked and leveled. And now, in the period of perestroika they level and level, and ditches on the sides, for draining do not match, the roads became worse! They do not have the technology, they are either alcoholics, or there is no one at all. Every spring it gets washed off. In the Soviet times this would not take place. It is just a dying village, there is nothing being built there. Gold miners needed it and don't seem to need it anymore.

Urbanization of traditional nature use

Indigenous peoples are often defined by their traditional uses of nature.[8] Although traditional nature use is usually viewed as the antithesis of the urban way of life, its sustainability is often directly related to successful communication by the representatives of indigenous peoples with the city. In particular, the main organizations working to defend the rights of the indigenous peoples

(Russian Association of Indigenous Peoples of the North, Siberia, and the Far East; the Batani Fund; and the Lyoravetlan Information Center) focus on traditional environmental management outside of cities. In general, they pay little attention to the problems of urban residents, aside from legislation to regulate contacts between indigenous peoples and city dwellers (Overland 2009).

At the individual level local hunters and trappers know that it is important to have relatives or friends who can negotiate deals for them in the city, where fur prices are much higher than in rural areas. Additionally, in the city there is a chance of getting orders for the well-known traditional medicines (*shilajit*, stone oil, herbal mixes) and tinctures (*Pantocrinum*) directly from the consumers who will pay much more than resellers in the village.

Often, leaders or representatives of *obshchinas* (legal entities with the rights for traditional use of natural resources, including hunting) live in county or regional centers or even in other cities, like Irkutsk. The 40-year-old chairwoman of one community explained her role:

> We have an "obshchina" there in Khamakar, the "Revival" headed by Tatiana Mongo. But she lives in Ulan-Ude, well, she organized [her relatives] too. ... I'm here (in Erbogachen, the county center) as the holder of stamps, money, for example, somewhere, some papers for something, I conduct communication like that.

QUESTION: That is, instead of the village council?

ANSWER: No, the village council is there (in Khamakar). But beyond that, there is also a social program, long-term development programs for the minority indigenous peoples. And now, we are participating in these programs.

In the Republic of Tyva, for example, programs that support reindeer herding are managed through established organizations with the accountants, program managers, and other administrative specialists required to operate as legal entities. It is easier to set up and register such organizations in the city than in rural areas. Specifically, a Republican Fund for Support of Reindeer Herding was established in Kyzyl. The main investors from Todzhinsk district are the Lunsin and Ak-Sug mining companies, as well as three gold mining companies that sponsor those programs. However, their involvement has actually increased rather than reduced the differences between traditional and industrial lifestyles. While the bureaucrats can report an increase in numbers as an achievement of the program's implementation, the reindeer herders will be the first impacted by the industrial activity:

> Money is allocated to our herders, who are grazing deer in the forest all year round. We also have a Municipal Unitary Enterprise, Udegen, where we have 18 reindeer herders. The number of heads of deer is growing here. This year we had 2,000, last year it was 1,200–1,300.[9]

Public hearings are held here, and reindeer herders roam the mountains and are the last ones to know about what is happening.[10]

In the neighboring Tofalaria, a reindeer herding base is registered in the administrative center, Nizhneudinsk, where all the related business is organized.

In Katanga district, advocacy on behalf of the indigenous peoples of Irkutsk oblast has been carried out by an advisor to the governor, N.G. Veysalova, who resides in the village of Erbogachen and who founded the Union for the Support of Indigenous Peoples of the North in Irkutsk oblast. Essentially, the Union became a foundation for the preservation of ethnic identity and the development of civic engagement, because along with the Evenki cultural center, they are the only non-governmental organization in the territory of the Katanga region. Based on their initiative, the Program for Support of Indigenous Peoples of Irkutsk oblast provided a mobile X-ray machine, attached to an all-terrain vehicle, for medical examinations in inaccessible settlements of Katanga district. They also established a Cultural Center of Indigenous Peoples of the Baikal region. However, while indigenous groups may be privileged in the autonomous regions, they are another minority group in the urban setting, along with Buryats, Belorussians, Poles, and so on, without any special rights or budgetary resources.

Often, the political activities of urban ethnic elites seem detached from the local needs:

> This is the type of organization where one can solve the hunting problems with a nod. And we have not seen it at all. It came during the election times and disappeared ever since. I have no faith in the work of the deputies, I am used to achieving everything myself.[11]

This lack of mutual understanding among some urban and rural Evenki was also noted by David Anderson, who conducted research in the Evenki autonomous region (Anderson 2000).

In order to consider the interests of indigenous peoples living in cities and suburbs, and to support traditional activities there, some cities and urban-type settlements of the North and the Far East were included in the list of places of traditional settlement. Such settlements do not exist in the Baikal region; however, even after leaving the village for the city, old-time hunters are trying to preserve their way of life, according to one hunter:

> They're coming in such a manner, to make it possible to go hunting, because in general, here remain those who are held back by hunting (they don't allow leaving, and postpone the move), not women. Dimka Arbatskii even managed to grab a land lot there. Sanka Griaz'nov there [in Severobaykalsk] got a land unit. Andrei Gudkov was on BAM in Taksimo, they have land there with brisk and winter stays and all. What I am saying is that in general, hunting is not

some fad; it is the meaning of life, it is impossible to live without it. Petya moved to Ust-Ilimsk, got a job, but Tribunskii noticed him here, he spoke with him, and immediately invited him to his place [he comes during the R&R period to Tokma, for hunting].

As the case of Alaska shows, indigenous people have been most successful utilizing traditional resources when cash income can be generated (Huskey 2010). In recent years, reindeer herders in Todzha and Katanga district have snowmobiles and boats in stock, while those who lost reindeer remain without any means of transport. The herders use reindeer when there is too little snow to use snowmobiles for transporting wood products or for travel. At the same time, the hunting sector is becoming more sophisticated. Hunters increasingly seek comfort and technology: they bring diesel generators, chainsaws, even satellite TVs to their hunting cabins, which then consume significant amounts of fuel.

Unlike their urban counterparts, hunters often review legal literature concerning their rights and other kinds of literature as well. Many hunting cabins I visited had glossy magazines devoted to the sport (*Okhota i rybolovstvo, Safari*, etc.). Given the low level of income of the hunters, such magazines seemed like luxury items to me. However, without any way to communicate with the outside world, except for the radio and magazines, the latter became a significant part of evening entertainment. Through legal and popular literature hunters learn how their lifestyle is perceived by the "outside" world and find ways to defend their interests.

With the use of photo and video cameras, as well as radiation dosimeters (in the case of metal ores), the ability of the hunters to monitor the activities of companies has slightly increased:

> The hunters drive around there. Now everything is easier, one can take a picture and show it to everyone. You can show the picture to them and say, why are you guys crossing the lines? They do not argue much, if they see that they were caught, they immediately remove everything and clean up. Well, and before being caught, they can get out of control.[12]

Another question is how this can really change the relationship between the indigenous peoples and the mining companies, as legally they do not have the rights to subsoil deposits.

Long-distance commutes and changes in the employment system

The long-distance commute method has become the main staffing strategy in the post-Soviet era. The emerging long-distance commute worker camps are a new type of settlement that are not included in the statistical studies and censuses, but they now involve an increasing number of people, sometimes equal to the permanent population of a particular area. These settlements gradually develop their own infrastructure, which is independent of existing local and regional facilities provided and controlled by government bodies. In 2012, a new airport was built in the area of Talakanskoye field to transport cargo and shift workers. Instead of using the services of the regional Angara Company, which handles more than half of the intra-regional flights, the mining companies used the services of UTair, which specializes in passenger and freight transportation in the oil and gas sector. The development of this airline coincides with the development of oil production. Specialists with experience and necessary skills often come from other oil regions (Tatarstan, Bashkortostan, Udmurtiya, and Western Siberia) and gradually circulate from one field to another, along with the mining companies.

Despite the proximity of remote villages with indigenous population to the fields, the expeditions prefer to announce employment opportunities in the cities and district centers and to provide transportation for workers from larger towns. Residents of these villages have to make their own arrangements to get to cities or regional centers instead of going directly from their villages.

Lee Huskey wrote that in the North there are only two kinds of economics: *international* – focused on large-scale resource extraction, and *local* – focused on local consumption of biological resources (Huskey 2010). In the international

Figure 6.1 Winter roads on frozen rivers (*zimniki*), Siberia's main transportation system.
Source: Photo courtesy of V. Lobchenko.

economy, the locals are more likely to participate in resource extractions that are close to their traditional activities, rather than those that utilize unknown technologies. The local economy is more characteristic of small settlements (less than 1,000 people). Representatives of indigenous peoples in Irkutsk oblast follow the same trends as those communities in Alaska, where the official unemployment rate is two times higher than that in the urban parts of the state; and the Canadian Arctic, where unemployment was about 40 percent among the indigenous peoples in 2000.

The indigenous peoples of Irkutsk oblast receive lower pension payments because of the low (two to three times less than in the region as a whole) proportion of people of retirement age, as well as a low proportion receiving official income from employment (less than a quarter of Tofalars and about a third of the Evenki have it – 41.9 percent for the region) (Sirina 2012). Overall, locals are more likely to be unemployed than non-locals, because the latter likely came to the region for skilled work, but in general, according to the census, the official unemployment rate among the minority indigenous peoples is almost two times higher than the average regional indicators (see Table 6.3). In addition, there is a high unemployment rate in the countryside as a whole. Table 6.3 indicates that the unemployment rate decreased among the urban population, including the representatives of minority indigenous peoples, in all regions except for Irkutsk oblast.

According to the analytical report on the socioeconomic situation in the Katanga region for the first half of 2013, the working-age population numbered 2,453 people while the working population in the district was estimated at 3,100 people, of whom 741 people were engaged in the extractive industry. This meant that at least 647 were non-locals who usually would come for the work in extractive industries. The average monthly salary for the studied period amounted to 73,761 rubles, while the regional average salary for the same period was 27,928.4 rubles (Ministry of Labor and Employment of Irkutsk oblast 2013). At the same time, in the region itself, population with incomes below the subsistence level constituted 810 people, and the unemployment rate in the labor force compared to the previous year increased by 15 percent (from 8.5 percent to 9.8 percent).

In Irkutsk oblast, residents of Irkutsk, Ust-Kut, and other cities along the railway are in a better position for working in rotating shifts. Some of the urban residents who moved to the city from the Northern regions retain their official registration (*propiska*) at a rural address in order to be able to use local benefits. For example, mining enterprise employees, even in the company's offices in the cities themselves, can obtain free hunting licenses, and take advantage of education quotas, exemptions from the army, etc. The same holds for Indigenous Peoples of the North. Living in Irkutsk, they have even more benefits for rotating shift work in their home area, as delivery of workers from Irkutsk is much better organized compared with other, closer, but less populated towns in the north of the oblast. Furthermore, local residents can switch the residence registration from a village, which has a negative reputation among the oil workers, in order to get a job as a resident of another settlement, with a more positive image.

Table 6.3 Population of private households in the age range 15–72, disaggregated by economic activity in the subjects of the Russian Federation

Regions	Population groups	Population size	Indicated economic activity	Including				% of those who indicated economic activity			
				Economically active	Of whom		Economically inactive	Economically active	Of whom		Economically inactive
					Occupied in the economy	Unemployed			Occupied in the economy	Unemployed	
Trans Baikal krai	Total	811853	768913	517346	448875	5353	251567	67.3	58.4	0.7	32.7
	Urban population	538477	502932	355426	321476	3394	147506	70.7	63.9	0.7	29.3
	Evenki	967	937	546	412	134	391	58.3	44.0	14.3	41.7
	Including urban	146	137	93	82	11	44	67.9	59.9	8.0	32.1
Irkutsk oblast	Total	1825500	1732619	1174621	1060262	114359	557998	67.8	61.2	6.6	32.2
	Urban population	1467770	1382474	963862	886376	77486	418612	69.7	64.1	5.6	30.3
	Tofalars	487	481	290	222	68	191	60.3	46.2	14.1	39.7
	Including urban	35	35	17	12	5	18	48.6	34.3	14.3	51.4
	Evenki	925	909	495	393	102	414	54.5	43.2	11.2	45.5
	Including urban	275	271	172	139	33	99	63.5	51.3	12.2	36.5
Republic of Buriatia	Total	708348	693581	446318	387787	58531	247263	64.3	55.9	8.4	35.7
	Urban population	417720	405968	276421	248985	27436	129547	68.1	61.3	6.8	31.9
	Soiots	2455	2438	1441	1112	329	997	59.1	45.6	13.5	40.9
	Including urban	177	173	110	92	18	63	63.6	53.2	10.4	36.4
	Evenki	2055	2028	1238	929	309	790	61.0	45.8	15.2	39.0
	Including urban	642	635	405	329	76	230	63.8	51.8	12.0	36.2

(continued)

Regions	Population groups	Population size	Indicated economic activity	Including Economically active	Of whom Occupied in the economy	Unemployed	Economically inactive	% of those who indicated economic activity Economically active	Of whom Occupied in the economy	Unemployed	Economically inactive
The Republic of Sakha (Yakutiya)	Total	714296	686808	493121	447867	45254	193687	71.8	65.2	6.6	28.2
	Urban population	473614	450573	334977	312654	22323	115596	74.3	69.4	5.0	25.7
	Evens (Lamuts)	10427	10148	6729	5557	1172	3419	66.3	54.8	11.5	33.7
	Including urban	3727	3608	2294	1988	306	1314	63.6	55.1	8.5	36.4
	Yukagirs	845	835	572	480	92	263	68.5	57.5	11.0	31.5
	Including urban	377	372	253	221	32	119	68.0	59.4	8.6	32.0
	Dolgans	1254	1239	818	689	129	421	66.0	55.6	10.4	34.0
	Including urban	203	192	117	100	17	75	60.9	52.1	8.9	39.1
	Chukchi	428	423	302	271	31	121	71.4	64.1	7.3	28.6
	Including urban	157	155	105	89	16	50	67.7	57.4	10.3	32.3
	Evenki	14334	13902	9492	8039	1453	4410	68.3	57.8	10.5	31.7
	Including urban	4014	3871	2563	2261	302	1308	66.2	58.4	7.8	33.8
Republic of Tyva	Total	208801	205741	123389	97176	26213	82352	60.0	47.2	12.7	40.0
	Urban population	112920	110675	68683	59503	9180	41992	62.1	53.8	8.3	37.9
	Tozhu Tuvans	1272	1254	753	480	273	511	59.6	38.0	21.6	40.4
	Including urban	2	2	1	1	–	1	50.0	50.0	–	50.0

Source: 2010 Census.

If locals have proven themselves to be good workers, they tend to follow the hiring company:

> My son worked in Yakutiya in the expedition for 13 years, now he went home. Now he settled here too, working in Naftabur. At first he worked there in Yakutiya, now here. Of course, it was hard to get a job, but due to the fact that he worked in Yakutiya, they looked in his labor book records, decided that they need such a person, and took him. Many locals here have a hard time getting a job. My youngest son was not able to get a job anywhere for a long time, he has a certificate of a feller. They went there too, about seven years ago. They worked for a season, and now they don't take them there. Because they say – we have Tatars working there, they have their own people.[13]

Currently, there are no reliable statistics on the number of rotation shift workers and on employment of the local people in the mining sector. Mining companies do not register in the local district. According to a report by the Katanga district administration, Verkhnechonskneftegaz employs 521 people, while, according to the company representative, the number of permanent and contract employees of the company in 2012 amounted to 3,500 people, with no expected reduction of the number of employees and the rate of production ("VCNG podvodit itogi yubileynogo goda" 2012). During the 2013 regional election campaign, the organization sent 188 workers, who were registered in Irkutsk oblast, to Erbogachen to participate in the election (according to the local election commissioner).

The situation has begun to change in recent years. Some expeditions take workers directly from the region and deliver them to the workplace via all-terrain vehicles or helicopters:

> If people are young, and have the strength – they go. They make good money ... within 60–70,000 ... (rubles) per month. They leave for six months straight. Then they come back and purchase cars. It wasn't like this here before, we didn't have this many cars. Just 5–6 years ago, outboard motors, you come to the parking lot, there were 2–3 foreign cars. And now in the parking lot there are only three of our Russian cars, from Soviet times. All the rest are foreign cars! In general, I'd say this: whoever wants to make money, can make it.[14]

A certain role here is played by personal acquaintance with the company, which helps with the employment of the person and his crew:

QUESTION: What do you think, is it possible to get a job in an oil company?

ANSWER: They certainly would not hire me anymore, but if I wanted to, then they would take me as a guard there, a security guard. But they do not decline the youth either. Now, if I turn to them with a request to hire someone, they would not refuse. They are even willing to arrange education at a college or school, and after graduation would hire. So, Nikita Vlasov asked for help with getting a job there. He entered without any problems.

QUESTION: And in the forest companies?

ANSWER: As an arrester – no problems of getting hired. Give them your guys for the rotating shifts. In this sense, they do not refuse. If there is a decent man, they will hire him with no problems.

QUESTION: Has anyone from Tokma worked for them?

ANSWER: Of course not! Nobody wants to. I offered them arresters. But the truth is they are on rotational basis, they have to live in Kirensk for month. Then they are taken home for a month. I offered this to my guys. It is the same for the oil workers, if a man is decent, not an alcoholic, for God's sake, we can arrange any type of employment. Well, they took a guard from Ikskoe. He arrived there, lived for a day and disappeared somewhere. You can understand them too, why would they need such workers.[15]

By repeating several times that a worker must be a "decent man," the respondent suggests certain qualities corresponding to the work ethic of these enterprises. This ethic is different from the standard in the local environment, which is predominantly hunters. Sometimes even the hunters recognize that they are unaccustomed to the standard working day:

The hunters have forgotten how to work like this; here they are their own bosses, they work when they feel like it. And this does not provide discipline, such an approach. And there they have to get used to the regime again, to the schedule, and it is challenging.[16]

In addition to labor discipline, representatives of mining companies indicate numerous problems with the recruitment of locals, related to poaching and theft. That aside, there are certain restrictions in employment related to the lack of required experience or qualifications among the locals:

It is not so easy to be hired by a good company quickly. Nowadays, in general, they do not hire without previous experience. Experience - where to get it from? One has to take care of his own employment.

QUESTION: And what about a wood-cutting job, at least?

ANSWER: Yes, you can get such a job, but you will need to have certificates that confirm your qualification to do it. Cutting wood is also a demanding job.

QUESTION: And what's the requirement there?

ANSWER: Certification as well! Certification that you have the right to cut timber. That do you have a level of qualification.[17]

The service industry can provide an additional source of income in the surrounding villages. In particular, some rotating shift workers visit the shops in Tokma village, but locals express discontent with the workers' use of heavy machinery as means of transportation. Also, local cafes have popped up along the winter road, where the main movement of shift workers and truckers takes place.

In the summer, in Tofalaria, delivery of technology for Zashihinskoe field comes from the Tulun region. In the winter, they use part of the way along the winter road – Nizhneudinsk-Alygdzher along the Uda River (Figure 6.1) – providing maintenance only for the section from the winter road to the mine. The road is maintained by an entrepreneur from Nizhneudinsk. The geologists hire laborers from the neighboring villages in Tulunsky district (Ikei, Ishidei, Arshan); it is easier to transport them alongside vehicle deliveries from Tulun. However, most specialists come from other cities:

> They generally attract professionals from Irkutsk oblast, there are people from Angarsk, Irkutsk, Usolye, Shelekhovo, Sayansk, and Svirsk.

QUESTION: And is there anyone from the Nizhneudinsk region?

ANSWER: I cannot say anything about Nizhneudinsk, but there are people from Tulun. Although we have an agreement that this year and last year we were ready to take local people to work as laborers, and different positions here. With Vladimir Alexandrovich we had a conversation, we do not deny our intentions ... We are geologists here, we can take on trades to dig ditches, shovel; we geologists keep our promise, we will take people on geological exploration. If there are certified experts with experience we will take them for machinery operation too.[18]

Local workers not only need transportation and lack qualifications, they also may be distracted by local social obligations and obligations to support the social infrastructure, as is demonstrated by the example in the Republic of Sakha:

> The shifts are rotational, so the prospectors finished their work and left. The prospectors, of course, sought to use the rotational shift system, so as not to pay for the infrastructure there, boilers, schools, kindergartens. All the best specialists have left, everyone left, and just the pensioners remain. And now they bring workers from Ukraine, at one time they brought them from

Cheboksary; there are basically people from Ukraine and Ural. Some people came from Irkutsk, but they are also fooled, they do not pay them as everywhere!

QUESTION: And do they come by summer?

ANSWER: In the spring they come to prepare, in March, in February, while cooperatives are still more or less in condition. In February delivery begins, and why in February, because it is necessary to prepare the equipment, and in the past the locals were making all the preparations in the winter, and of course they had to pay them and everything. And back then they would give the first gold by May 9. This suggests that all the equipment was prepared. I myself participated in the equipment setting. We were sent there. What benefits they got out of this, I don't know. They got some benefits for themselves certainly. And so, basically they drove all the people out. Some people went on to work on diamonds; others went to cooperatives, that's all. Everyone from Neryungri left. The salary is too small. Well, they put pressure on the locals, to drive everyone out. Voluntarily or whatever. For example, someone comes back from diamond prospects; the miners do not hire this person, saying: "I need someone to work with gold, while you worked on diamonds." You have to pay the locals, they know better already. And strangers, they are happy with pennies, they come from the villages … They do not hire the locals, because the locals begin to snap immediately. Well, they just take a few bums. They take them to show that they hire locals. Well, they gather up all sorts of bums, cripples, and alcoholics, they work for a month or for a half, all they need is food. They eat and that's all. In exchange, they will make a mark that they hired such and such. Yes, they take the local Evenki, to crack ice, to dredge thaw, but they take them only for the report. They can take up to 50 people. And then they start drinking, get kicked out, sometimes they don't pay them.

QUESTION: And how many strangers do they hire?

ANSWER: Well, up to 300 people, for example. And then some move on, some go to another job. Some go home.[19]

According to the local press, in Todzhinsk district (Kozhuun): "The number of employees in the Lunsin Company amounted to 1,412 people on June 1, 2013. Among them there are 677 Russian citizens, residents of Tuva, including 75 residents of Todzha … from the moment the Company starts to receive profit (i.e., from the moment the mining and processing plant is at full capacity), the Lunsin Company will participate in the joint construction of the following object on the territory of the Todzha district: a bridge over the rivers of Tora-Khem and the Yenisei, a kindergarten in the regional center with 250 person capacity, and an elementary school, as well as equipping the district hospital with modern medical equipment" ("Budushchee sozdaetsya segodnya" 2013). According to the materials of the local administration, there are 310 Russian workers in the field, of whom 55 are residents of the district (Kozhuun) (20 drivers, 8 other skilled workers [experts, doctors, economists, engineers, etc.], and the rest are employed as unskilled laborers). According to the commitments, the number of foreign workers (from China) shall not exceed 50 percent. There is a contract with the district hospital signed for 91,000 rubles; half of the amount goes to salaries, and the other half is spent on fuel for delivery of physicians who take shifts once a week.

Conclusion

The urbanization of the indigenous peoples of the northeast Siberia is a complex and ambiguous process. The Soviet legacy continues to have great influence through the educational system (universities specializing in training representatives of indigenous peoples, and admissions quotas), and in the structure of transport routes (via regional hub). In addition, the Soviet system of resettlement and the importance of administrative ties with the municipal and regional centers in the allocation of resources, even of traditional nature, has been preserved and even strengthened. Due to these factors, representatives of the indigenous peoples tend to migrate to the educational, administrative, and regional centers, rather than neighboring cities. Russian old-timers, however, tend to migrate to neighboring towns.

Other significant changes in urbanization are associated with the development of mineral and fuel resources. When mining companies become involved, the process of modernization of the indigenous peoples gets accelerated. Unlike traditional livelihood systems, modern lifestyles require considerable material supplies, primarily fuel. Although highly volatile and dependent on the external resources, it is perceived as a more "civilized" system.

Indigenous peoples are becoming more involved in the modern urbanized and globalized world through market and bureaucratic communication, supply chain and life support, family and professional networking, education and transport, and electronic media. However, the degree of involvement will vary depending on specific cases. In contrast to the autonomous regions, the studied communities have fewer opportunities for being more visible in the regional centers, where the distribution of resources and power takes place.

As in many other studies of migrations, modern technologies allow individuals to preserve and maintain long-term ties with their place of birth. Further study is needed of how these links are used as a resource for both rural and urban residents, offsetting weakness and discontinuity of information links between the mining companies, the government bodies, and mining areas – areas of traditional indigenous settlement.

Notes

1 In order for a settlement to be classified as a city, it must have a population of at least 12,000 people, with the proviso that 58 percent of the population consists of workers or servers. Settlements qualifying as towns require a population of no less than 3 million people, of whom 85 percent should be employed outside of the agricultural sector.

2 Evenki woman talking about her children, who moved to Ust-Kut.

3 Woman, 55 years old, Tofalaria.

4 Woman, 55 years old, Tofalaria.

5 Woman, 50 years old.

6 Interview with a local hunter.

7 Woman, Tokma village, 58 years old.

8 Non-numerous indigenous peoples of the Russian Federation (hereinafter, indigenous minorities) are defined as the peoples living in the territories of traditional settlement of their ancestors, preserving the traditional way of life and traditional livelihoods and craftsmanship, with populations in the Russian Federation of less than 50 thousand people that identify themselves as separate ethnic communities (Federal Law of April 30, 1999, No. 82-FZ "On guarantees of the Rights of Minority Indigenous Peoples of the Russian Federation," with amendments on August 22, 2004, and June 26, 2007).

9 Head of Todzhinsk district (Kozhuun).

10 Woman, 40 years old, Toora-Khem village.

11 Interview with Evenki woman, Republic of Buryatiya.

12 Male, 45 years old, community representative.

13 Woman, 63 years old.

14 Man, 45 years old.

15 Interview with community leader.

16 Interview with local hunter.

17 Man, about 30 years old.

18 Interview with a geologist.

19 Interviews with pensioners living in Irkutsk, who go home to the Republic of Sakha (Yakutiya) for the hunting season.

Part III
A growing multiculturalism

7 Social dynamics and sustainability of BAM communities

Migration, competition for resources, and intergroup relations[1]

Olga Povoroznyuk

The BAM region refers to the territories situated along or connected to the Baikal–Amur Mainline. This is the most important Northern transportation route, a railroad system transecting six federal subjects of Eastern Siberia and the Russian Far East and linking Eurasian countries with East Asia. The logistical and geopolitical importance of the transportation route in this sparsely populated Northern region, as well as its large mineral deposits of gold, copper, rare metals, and coal, prompted the railroad's construction in the Soviet period. Recent socioeconomic trends show a renewed interest in the BAM, primarily to facilitate the extraction and transportation of mineral resources to Asian markets.

The cities and towns located along the mainline were originally intended to be temporary settlements or railroad stations in proximity to the existing indigenous and mixed communities. Currently the BAM network encompasses 210 railway stations, some of which gave rise to the large cities of Ust'-Kut, Severobaykal'sk, and Tynda; towns like Taksimo, Novaya Chara, and Khani; and a number of smaller settlements (see Map 7.1) built by labor migrants from particular Soviet republics, regions, or cities. The settlements' man-made environment – such as the architectural design of railway stations, city planning, and street names – still reflects the cultural or ethnic particularities of the builders. From a socioeconomic point of view, BAM urban settlements are typical single-industry towns dependent upon the railroad and extractive companies.

The social and ethnic fabric of BAM communities is woven of three main population categories: indigenous people (*aborigeny*); BAM builders (*bamovtsy*); and industrial shift workers and more recent labor migrants from the post-Soviet space (*priezzhie*). In this region, Evenki people have lived alongside Russian settlers and people of mixed origin (*gurany*) for centuries, whereas BAM builders and new migrants have been arriving since the late Soviet period. Relations among the main groups have varied from peaceful coexistence to competition and social tensions. Currently, life along the BAM implies existing within a fluid system of stakeholders, including state authorities, private industrial companies, employees of nongovernmental organizations (NGOs), and indigenous enterprises (*obshchinas*) that represent the interests of different groups. In BAM settlements, where labor migrants constitute the majority population, being part of a local community is a valuable asset in local identity politics (Sokolovskii 2012).

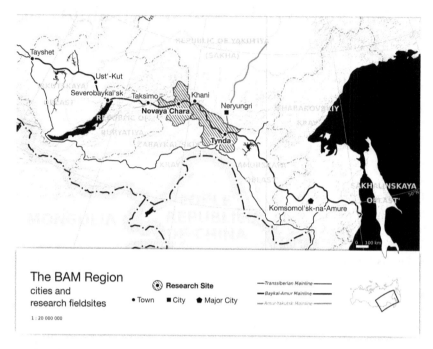

Map 7.1 The BAM region, cities and field sites.

Source: Map data © OpenStreetMap contributors.

The chapter draws on the field data collected in the Northern parts of two federal subjects lying within the BAM zone – Amurskaya oblast (Tynda and Tyndinskii rayon), and Zabaykal'skii krai (Novaya Chara and Kalarskii rayon) – in September–October 2013, with a focus on two urban communities – Tynda (population ca. 36,000), the hub city and the "capital" of the BAM, and Novaya Chara, a medium-size town of about 4,300. This chapter explores the social dynamics and sustainability prospects of BAM communities in the late Soviet and post-Soviet periods. I claim that the BAM has served not only as an important transportation route, but also as an agent of social change connected with migration and the formation of a culturally, ethnically, and socially diverse local population.

Further, I examine the sustainability prospects of the local communities dependent on the railroad and extraction industry, applying the concepts of the single-industry town and the resource curse that reflect the unsustainable development path of many countries and regions endowed with natural resource wealth (Kronnenberg 2004). This phenomenon is also characteristic of the current industrial development of the BAM region as an emerging Northern resource frontier. Finally, I will demonstrate that the strongest "pillar" for the sustainable development (Colantonio 2007) of BAM communities is social and human capital accumulated in the process of successful integration of newcomers, nation-building, and community identity construction. Cultural and ethnic tolerance carries the most significant potential for creating a favorable social environment, while traditional industries and the emerging field of ethnic tourism present attractive prospects for the sustainability of BAM communities.

Industrialization history of the BAM region

The BAM region (or the BAM zone) is a term used to describe the territories adjacent to and dependent on the infrastructure of the Baikal-Amur Mainline. BAM's legacy begins in the late nineteenth century. With the outbreak of World War I, the tsarist government built a railroad at the southern shore of Lake Baikal in an attempt to ensure the geopolitical security of the Russian Far East and East Siberia against China. The next ancestor of the contemporary BAM was the railroad stretching from Komsomol'sk-na-Amure to Sovetskaya Gavan' in Khabarovskii krai, built between 1932 and 1953 by labor camp inmates, military personnel, and prisoners of war (Mote 2003). That project was abandoned after Stalin's death in 1953, and the idea of restarting the BAM construction gained official favor only in the Brezhnev era, nearly two decades later.

"The third BAM" represented a grandiose engineering endeavor and the last megalomaniac Communist industrial project exploiting the USSR's vast natural resources for propagandistic and economic reasons. Moscow hoped that a completed BAM would bolster collective faith in the command-administrative system and serve as the prototype for further conquests of the Soviet Union's vast and resource-rich northeastern frontier in the twenty-first century (Ward 2009, 2–5). In 1974, the Komsomol, the Communist Party's youth organization, announced the beginning of BAM construction and a youth labor mobilization campaign. Soviet propaganda urged young people to rally together and build BAM in the spirit of "self-sacrifice" and "fraternal cooperation" for the sake of "social strengthening" in the remote corners of the USSR (Brezhnev 1993).

The majority of the mainline was built between 1972 and 1984, although some sections were put into operation as late as in 2003. Due to its high construction and maintenance costs and the fact that the railroad has never operated to its full capacity, the BAM has been considered an unprofitable enterprise. In the 1990s, these circumstances resulted in public criticism of the BAM project, the loss of the project's social prestige, its absence from the public spotlight, and further decline. In 1997, the BAM network was transferred from the state-owned Baykalo-Amurskaya Zheleznaya Doroga company to Rossiiskie Zheleznye Dorogi (RZhD), currently Russia's largest state railroad company.

The present day BAM is approximately 4,300 kilometers (2,600 miles) long, with its main branch, the Amur-Yakutsk Mainline (AYaM), constituting 1,200 kilometers (746 miles). The Mainline crosses the Northern districts of six federal subjects – Irkutskaya oblast, the Republic of Buryatiya, Zabaykal'skii krai, in East Siberia, and the Republic of Sakha (Yakutiya), Amurskaya oblast, and Khabarovskii krai in the Russian Far East. With its existing and projected side-tracks leading to mineral deposits and connecting remote settlements with administrative centers, the railroad provides a reliable transportation network for people, goods, and resources.

The mineral resources of the BAM region include the largest coal deposits (Neryungri field in South Yakutiya, the Apsat deposit in Northern Zabaykal'skii krai, the Bureya deposit in Khabarovskii krai), oil and gas deposits (Markovskoe, Yaraktinskoe, Ayanskoe) in Irkutskaya oblast, nonferrous and rare metals deposits (Udokan copper deposit in Northern Zabaikal'skii krai, the Kholodninskoe and Ozernoe lead and zinc deposits in Buryatiya), as well as numerous other ferrous metal deposits scattered across the region. In the 1980s, Moscow drafted plans to establish several regional industrial clusters, such as the Udokan mining and processing plant, as well as coal mining centers in South Yakutiya and Khabarovskii krai, which would have resembled the existing industrial centers in Ust'-Kut and Komsomol'sk-na-Amure (Aganbegian et al. 1984, 9–11). These state plans, however, have not been realized due to the economic collapse.

BAM communities

Settlements located in the BAM region, in the general sense, vary from cities to small indigenous and mixed villages. However, in this chapter I focus on BAM communities (*bamovskie poselki*) located along the mainline or in immediate proximity to it. These communities emerged as railway stations and temporary industrial settlements, some of which have grown into towns and cities. Remarkably, each BAM settlement with the adjacent infrastructure was usually built by a "patronage team" (*shefskaia brigada*) from a certain city or region in Central Russia or from a Soviet republic. The BAM was considered to be "the incarnation of friendship and cooperation between all peoples in the USSR" (Brezhnev 1993, 92). For example, the city of Tynda, where BAM and AYaM crisscross, was built by Muscovites, and the city of Severobaykals'k by Leningraders. Smaller towns and settlements along the railroad were built by Kazakhs (Novaya Chara), Uzbeks (Kuanda), Turkmens (Larba), Georgians (Ikab'iya, Niya), Armenians (Yanchukan, Tayura), Azeris (Ul'kan), Moldovans (Alonka), Estonians (Kichera), Latvians and Belorussians (Taksimo), Lithuanians (Novyi Uoyan), Tajiks (Soloni), and so on.

Currently, the BAM encompasses over 200 stations and traverses 65 villages and towns. According to the 2010 federal census, the largest cities of the BAM include Tayshet (35,485 people) and Ust'-Kut (45,375 people) in Irkutskaya oblast, Severobaykal'sk (24,929 people) in Buryatiya, Tynda (36,275 people) in Amurskaya oblast, Neryungri (61,747 people) in Yakutiya, and Komsomol'sk-na-Amure (263,906 people) in Khabarovskii krai.[2] However, the typical BAM settlement is a town of a smaller scale. There are 14 urban BAM communities in Tyndinskii rayon, Amurskaya oblast, with the population ranging between 243 residents in Amosovskii and 3,029 in Yankan.[3] There are four BAM communities (three rural and one urban) in Northern Zabaykal'skii krai. In 2010, Novaya Chara, the only urban community of Kalarskii rayon, had a population of 4,315 residents.[4] The population size of BAM communities peaked in the mid-1980s and rapidly declined during the socioeconomic crisis and mass exodus of the population from the North in the 1990s (Heleniak 2010). Currently, BAM communities continue to lose their residents, with the annual, well-documented out-migration from Siberia and the Far East. For example, in 2012, Tynda lost 733 residents due to out-migration.[5]

Most BAM communities resemble typical single-industry towns that depend on the functioning railroad and developing extractive industries. According to economic criteria, a monotown (*monogorod*) is a town that has one or more enterprises functioning as a single production cluster, which employs over 25 percent of the economically active population, accounts for more than 50 percent of overall industrial production, and, optionally, for over 20 percent of all organizational taxes and revenues to the municipal budget (Animitsa 2010, 9–10). In Russia, monotowns are a Soviet legacy: their foundation and development was a means of adaptation and territorial organization of workforce in the USSR's geopolitical, economic, geographic, and climatic context. The collapse of the planned economy caused a decline or complete closure of backbone enterprises, creating numerous socioeconomic problems for the residents of such towns. Post-Soviet monotowns are characterized by a homogeneous occupational structure, high levels of unemployment, underdeveloped social institutions, and insufficient cultural and educational opportunities (15). In addition to economic dependencies, they experience a crisis of social and cultural self-determination: while the residents are attached to their communities, they tend to refrain from participating in public life and local politics.[6]

In Northern Amurskaya oblast, transportation is the major industry since the mainline plays a paramount logistical role. In Tyndinskii rayon, the railroad enterprises employ approximately 20 percent of the local economically active population.[7] In BAM communities, the overwhelming majority of the local labor force work for railroad service and maintenance companies, while many others are employed in public organizations or look for shift-based jobs in large-scale mining companies operating in the same or neighboring districts and regions. In contrast to Soviet times, when the intelligentsia from all over the USSR flew to the BAM zone, current public institutions, like the district hospital in Tynda, lack specialists because young doctors are not willing to stay in the city, let alone smaller BAM towns. At the same time, Tyndinskii rayon has high unemployment levels. Low-skilled job seekers from BAM towns in Northern Amurskaya oblast and Zabaykal'skii krai resort to shift work for the Petropavlovsk mining and processing group.

The physical environment of BAM towns, including architecture, place names, street layout, and other details, reflects the cultural or ethnic particularities of the builders. Thus, the railway station in Novaya Chara in Zabaykal'skii krai resembles a Kazakh yurt (Figure 7.1), and the street names (i.e., Arbat, Krasnaya Presnya) and high-rise buildings in Tynda resemble those of Moscow (Figure 7.2). The social infrastructure of the BAM urban communities includes necessary facilities such as administrative buildings, kindergartens, schools, hospitals, fire stations, and shops, depending on the size of a community. Cities like Tynda, "the capital of the BAM," have a well-developed trade and service sector, including large shopping centers, restaurants, and fitness clubs, as well as museums, theatres, exhibition and concert halls, churches, and monuments commemorating construction of the BAM (Figure 7.3). However, many ambitious urban construction and development projects announced in the BAM

Figure 7.1 Railway station in Novaia Chara, Zabaikal'skii region (photo by the author).

Figure 7.2 High-rise apartment buildings in Tynda, Amurskaya Province (photo by the author).

Figure 7.3 Installation commemorating the 40th anniversary of the construction of the BAM in Tynda, Amurskaya Province (photo by the author).

heyday have not been implemented due to the recent economic crisis. In Chara, several foundations for unfinished apartment buildings dot the contemporary cityscape, while the decaying foundation of a shoemaking factory in Tynda reminds the city's residents and visitors of Soviet-era construction plans.

In 2006, RZhD Company started to transfer housing, utilities, and other social infrastructure in BAM communities to district and local municipalities, whose budgets cannot afford this burden. Over 50 percent of the housing stock in Tynda and Chara is decrepit. The same is true for the roads connecting the district centers to other BAM towns and villages. The significance of such poorly maintained roads became evident when new restrictions were introduced barring the use of BAM work trains to transport civilians.

Currently, federal investments in community development in the BAM region are drying up, and local revenue related to the railroad and resource extraction are not sufficient to fill the budget gap. In order to compensate for the high construction and maintenance costs of existing infrastructure, local authorities often appeal to extractive companies in the region. However, their support is officially recognized as voluntary, as federal administrators do not anticipate any regular revenues from mineral extraction flowing to local budgets. Thus, the current social programs of extractive companies in Kalarskii rayon are limited to occasional one-time funding of social and cultural events and selected construction and renovation projects. In Tyndinskii rayon, the Petropavlovsk and Priisk Solov'evskii companies make more visible investments in the social infrastructure of BAM communities.

The population and identities

The population of BAM towns is comprised of three categories of people: the indigenous population (*aborigeny*, *korennye*), including Evenki and Russian Old Settlers of mixed background (*gurany*); BAM builders (*bamovtsy*); and migrants (*priezzhie*) (Bulaev 1998). These labels draw on a specific genealogy of ethnic and nation-building policies and classifications dating to Soviet times. Currently, these and other names are used by representatives of different groups and stakeholders in their claims to belong to local communities.

Aborigeny

The Evenki, a Tungus-speaking minority group, are the indigenous population of the region. They generally refer to themselves as "aboriginals" (*aborigeny*), whereas other groups may also call them "indigenes" (*korennye*). Evenki nomadic reindeer herders and hunters were gradually sedentarized through the Soviet collectivization campaign, agricultural reforms, and "cultural construction" carried out among the indigenous peoples of the North. The shift-work method, introduced to so-called traditional activities (herding, hunting, and fishing) ostensibly to increase their productivity, in reality resulted in gender-based socio-professional distortions within indigenous communities. Whereas indigenous men continued working in the taiga, women with children settled in villages and became employed in the administrative or public sector. Indigenous enterprises (*obshchinas*), which appeared after the reorganization of collective farms (*kolkhozes*), now employ mostly male herders, whereas few reindeer herding families still pursue the traditional way of life (Povoroznyuk 2011) (Figure 7.4).

According to the 2010 All-Russian Census, 501 Evenki lived in Kalarskii rayon, Zabaykal'skii krai.[8] The main places of their residence located along the railroad include the BAM settlements of Novaya Chara (17), Chara (53), Ikab'ya (19), and Kuanda (29) and adjacent indigenous village Chapo-Ologo (141). The remaining Evenki population lives in the villages of Kyust'-Kemda (56 persons) and Srednii Kalar (39 persons).[9] In Tyndinskii rayon, Amurskaya oblast, 810 Evenki live in the villages of Ust'-Urkima (221), Ust'-Nyukzha (401), and Pervomayskoe (188) located along the BAM.[10]

Documents provided by the Administration of Kalarskii rayon in September 2013 show that the majority of Evenki lead a sedentary life working in state-funded organizations, while only 41 are registered as reindeer herders and 99 individuals as hunters, respectively. The lands officially designated for herding and hunting are allocated to six Evenki enterprises, according to land lease agreements. However, de facto, the number of unregistered family-based units pursuing traditional economic activities is bigger. There are 17 registered enterprises

Figure 7.4 Evenki reindeer herders' family in their village house, Chapo-Ologo, Zabaikal'skii region (photo by the author).

(*obshchinas*) involving 63 persons and owing 4,797 reindeer in Tyndinskii rayon.[11] In both districts, Evenki are also entering the emerging fields of ethno-tourism, souvenir production, and reindeer antler procurement.

In recent years, the notorious social problems of unemployment, low living standards, and alcoholism among the indigenous population have been aggravated by the withdrawal of internationally recognized indigenous rights, particularly in the sphere of land use, from federal and regional legislation and cuts in state support (Yakel 2012). While the Russian Federation formally participates in the Second International Decade of the World's Indigenous Peoples and has a special state program for socioeconomic development of the indigenous peoples of the North, in practice, the major funds allocated under this program go to the (re)construction of social infrastructure, facilities, and housing in indigenous villages.[12]

Despite these facts, Evenki are the most stable population of the BAM region. They are characterized by low levels of social and geographical mobility. Their routes include taiga camps, indigenous villages, and the nearby BAM settlements, usually within the same federal subject or within the BAM region. Such strong attachment to the place of birth and residence are predetermined by kinship and family ties, on the one hand, and the lack of educational and employment opportunities in other places, on the other. Thus, when juxtaposing themselves with other groups, Evenki use the term "aboriginals" (*aborigeny*) in order to highlight their rootedness in the region. At the same time, *bamovsty* and other residents, who were born or spent most of their life in the region, claim to belong to the same group of *aborigeny*, while referring to Evenki as indigenes (*korennye*) in the contexts when they wish to stress their cultural difference.

Bamovtsy

In the 1970s and 1980s, the construction of the Baikal-Amur Mainline attracted labor migrants from other Russian regions and other former Soviet republics (Belkin and Sheregi 1985). Generally referred to as BAM builders (*bamovtsy*), about 500,000 temporary BAM workers were lured to the region by the Communist party's youth organization, the *Komsomol* (Ward 2009). The labor force recruited to build the BAM was largely comprised of young, educated, and skilled men, who initially came to work on a contract, but often married and settled in the region. One-third of the BAM builders arrived from different parts of Russia, one-fifth from Central Asia, particularly Kazakhstan, and the remaining part from Belorussia, the Baltics, and the Caucasus (Argudyaeva 1988, 9–11). In 1984, the European part of Russia (19 percent), the Far East (18 percent), and Ukraine (15 percent) were the main home regions of the BAM migrants, whereas indigenous people accounted only for 1 percent of the local population (see Table 7.1).

Table 7.1 Composition of residents of the BAM zone (percent) by home regions

Home region	1981	1984
Far East	4.5	17.8
East Siberia	15.8	14.1
West Siberia	7.7	5.0
Ural	6.4	8.5
European part of Russia	15.3	19.4
Ukraine	21	15.2
Moldova	5.3	13
Belorussia	4.2	2.5
Baltics	5.5	1
Kazakhstan	4.3	7.8
Central Asia	5.5	6
Caucasus and Transcaucasia	1.2	0.9
Indigenous population of the region	3.3	1.2

Source: Based on Argudiaeva 1988, 1.

Young BAM builders were motivated by Communist ideology and the romanticism of the Komsomol youth movement, a characteristic of other Soviet large-scale industrial projects (Rozhanskii 2002). The image of the BAM as "the building site of the century" as well as other propagandistic slogans and clichés were used to lure young workers from across the USSR to the railroad construction and, later, to develop solidarity among *bamovtsy*. Prior to enrollment in a BAM construction brigade, a specialist was supposed to meet certain educational and professional requirements and to demonstrate his or her motivation and compliance with Communist ideals.

Yet, during late Soviet socialism, builders were also attracted by the lucrative material benefits. According to the contracts that workers concluded in their home republics and regions, the state provided them with apartments and cars after several years of work, as well as high salaries and other social benefits. The BAM builders also enjoyed access to goods and commodities regularly supplied to the region but unavailable elsewhere in the country. As a result, a contract at the BAM often yielded a substantial amount of income in a relatively short period of time. Such opportunities attracted not only specialists, but also fortune-seekers – short-term contractors, and, since the 1990s, individual entrepreneurs and dealers.

The former builders that I interviewed fondly recalled the sense of solidarity and communal feeling among *bamovtsy*. Many informants refer to the period of BAM construction as the happiest time of their lives (Bogdanova 2013), when people were friendly, helpful, and supportive of each other. In addition to the ideology and material benefits, the sense of unity and belonging was achieved through social factors – a mostly homogeneous age, educational, and professional profile that facilitated social networks. For example, the neighborhood settlement patterns, wherein colleagues working in the same organization or construction brigade also became neighbors in their apartment buildings and, thus, spent time together both at work and at home, strengthened friendly ties. The ideological and social underpinning of the BAM railroad and related settlements' construction gave the builders a sense of fulfillment, contributing to the very creation of the place and the following attachment to its social and built environment (cf. Bolotova and Stammer 2010, 208).

> While people in the country had nothing to eat or drink, there was everything here; there was a little communism here. We could easily buy what people couldn't afford in 30–40 years of intensive work. In three years, we could buy a car. We had free money, so we could go on vacation. We lived a rather wealthy life. In a certain period of time, we developed an "affection for the North," some kind of attachment: once you came here, it's hard to leave. There is a special type of people here – open-minded, kind, hospitable, ready for selfless help. The North engages you, the North makes you a hostage. We make our mind to leave, but then change our point of view. We got used to here, we feel comfortable and cozy.[13]

In the 1990s, the BAM region witnessed a large-scale out-migration of the non-local and non-indigenous population. The socioeconomic crisis of the 1990s drove the majority of BAM builders from the North. Local authorities estimate that approximately one-third and one-half of the *bamovtsy* population relocated to Kalaraskii rayon and Tyndinskii rayon, respectively. These were the people who had participated in the construction and early use of the railroad. Currently, the *bamovtsy* category has a broader interpretation. It includes an informal community of "the children of the BAM," the second generation of BAM builders who spent their childhood and, sometimes, part of their adult life, in the region. This group name has also been self-ascribed to the specialists and entrepreneurs who "came to work at the BAM" in the 1970s–1990s, but did not directly contribute to the railroad construction and maintenance process itself.

In the 1990s, the BAM project became an object of open criticism and public amnesia due to its unprofitability in the context of the socioeconomic crisis and ideological turn. However, a decade later, the BAM was again regarded as a unique technological, socioeconomic, and ideological endeavor. In fact, the BAM turned out to be a testing ground for Soviet ideological, nation-building, and economic policies. The BAM legacy was reflected in the local folklore and art exhibited in museums, and became part of *bamovtsy* life stories and the recently rehabilitated social memory of the region.

Priezzhie

At the local level, the term *priezzhie* (newcomers) is usually associated with recent migrations and migrants, who constitute an insignificant proportion of residents, not enough to compensate for the departing population. In recent decades, there have been two distinct categories of *priezzhie*: (1) long-term or permanent migrants from other regions and districts and the post-Soviet states and (2) shift workers (*vakhtoviki*) outsourced by extractive companies from the neighboring districts and regions and, more rarely, from other parts of Russia (see Chapter 5). In 2012, the majority of Russian migrants (10,507 out of 23,245 people who moved to Zabaykal'skii krai) indicated family as their main cause of migration, whereas for most international migrants (474 out of 814 people) a job was the main factor.[14]

Migrants from other parts of Russia usually find employment in public sector, especially with RZhD, while some of them start their own business in trade and services. International labor migrants from the former USSR arrive at BAM settlements mostly from Uzbekistan, Tajikistan, Kyrgyzstan, Armenia, and Ukraine. This is the smallest and most diverse and scattered group, who tend to be self-employed, on a semi-legal basis, in trade, agriculture, and services. However, those who manage to acquire Russian citizenship become eligible for more attractive job opportunities in both the public and private spheres. In Novaya Chara, Kalarskii rayon, several such employees work as track inspectors for the East Siberian Branch of RZhD. In Tynda, some migrants from Central Asia and the Caucasus keep their own smaller farmsteads at the city outskirts; others open grocery and flower shops and restaurants in the city center. They are also visible at open-air flea markets and in the main shopping malls.

The growing number of international migrants from the post-Soviet space has recently come to the attention of local and regional authorities. While federal legislation and the state strategy for ethnic policy regulate the legal status of migrants, the social pathways for their integration are still developing. In 2013, a center for sociocultural assimilation opened in Chita. It plans to offer courses in the Russian language and migration laws and to provide social and psychological support to migrants' families. However, in the Northern districts of the BAM region, such centers are still lacking; therefore, social networks, including familial and friendship ties, traditionally help migrants adapt to the new social and cultural environment.

The mining companies currently in the region mostly recruit labor from other parts of Russia using fly-in/fly-out shift work (*vakhtovyi metod*). According to some estimates, Sibirskaya Ugol'naya Energeticheskaya Kompaniya (SUEK), which is developing the Apsat coal deposit, employs over 200 shift workers (*vakhtoviki*), who previously worked for its subsidiary company in southern Zabaikal'skii krai. BGK Company, operating at the Udokan copper deposit, has only a few employees from the local population, while the majority of its labor force (approximately 120 qualified mineral engineers and other specialists) are shift workers from other parts of the region (Figure 7.5). Experts from the Ministry of Natural Resources in Chita argue that the fly-in/fly-out method has proved effective in the Northern conditions and will be increasingly used in mining in future,[15] in contrast to the Soviet era, when labor was recruited by building large, permanent settlements (Heleniak 2010, 33). Thus, *vakhtoviki* are a growing socio-professional group of *priezzhie* in the BAM region.

Figure 7.5 Shift workers at the experimental plant, Udokan deposit, Zabaikal'skii region (photo by the author).

Social dynamics: migrations, ethnicity, and intergroup relations

As mentioned above, since the 1990s, the population of the BAM region has been steadily decreasing. Figure 7.6 shows the data for Zabaykal'skii krai and Amurskaya oblast. In Kalarskii krai, the recent migration loss fluctuated between 91 people in 2006 and 239 people in 2012. Among 390 migrants who left the district, there were 306 interregional, 80 interdistrict, and four international migrants in 2012 (see Figure 7.7). The majority of interregional migrants left Zabaykal'skii krai for other parts of Siberia and the Far East, South Russia, and the cities of Moscow and St. Petersburg.[16] Amurskaya oblast, located farther north, has faced an even higher migration loss. The out-migration from Tyndinskii rayon took away 540 people in 2011 and 529 in 2012,[17] whereas the outflow from Tynda municipality was 787 and 733 persons in 2011 and 2012, respectively.[18]

The changing socioeconomic situation in Russia has affected migration as well. The population leaving BAM has been partially substituted by a growing wave of post-Soviet migrants from other Russian regions (interregional migrants), other parts of the region (interdistrict migrants), and, to a lesser degree, international labor migrants, arriving, mostly, from the post-Soviet space. Among 142 migrants who arrived at Kalarskii rayon in 2012, there were 88 interregional, 48 interdistrict, and six international migrants.[19] The migration statistics for Zabaykal'skii krai reflect the general trends for its Northern BAM region: the majority of interregional migrants arrive from the neighboring federal subjects and the Far East (11–13). As for international migration to Zabaikal'skii krai, Uzbekistan, Tajikistan, Kyrgyzstan, Armenia, and Azerbaijan account for the overwhelming majority of migrants from the post-Soviet states, whose number

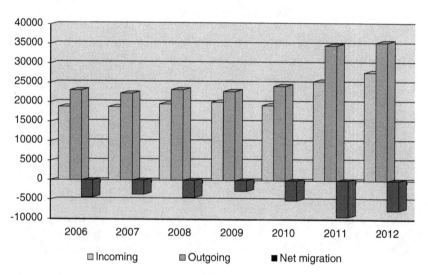

Figure 7.6 Migration dynamics in Zabaikal'skii region (number of residents).

Based on: Migratsiia 2013, 31.

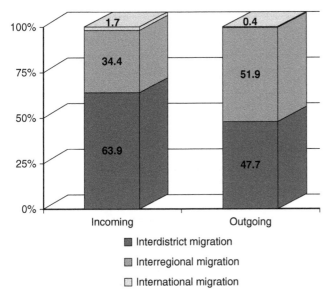

Figure 7.7 Distribution of migrants in the district of the Far North, Zabaikal'skii region (percent) in 2012.

Based on: Migratsiia 2013, 33.

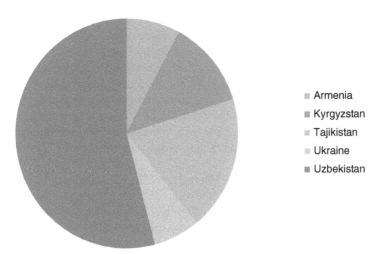

Figure 7.8 The structure of the international migration from the post-Soviet countries to Zabaikal'skii region in 2013.

Based on: Svedeniia 2013.

has been steadily growing since the early 2000s (see Figures 7.8 and 7.9).[20] Most of these migrants settle in the region's capital city, Chita, while some move up North to the BAM area.

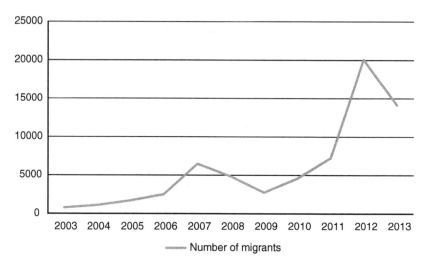

Figure 7.9 Dynamics of migrations from the post-Soviet countries to Zabaikal'skii
region in 2003–2013.

Based on: Svedeniia 2013.

The mass out-migration from the Far North in the 1990s can be explained by
the "surplus workforce" phenomenon reaching back to Soviet industrialization
practices. In fact, during the post-Soviet socioeconomic crisis, loss of jobs was
an important driving force behind out-migration from the North. However, since
the 2000s, unemployment has been coupled with other social, economic, and
environmental hardships experienced by Northern residents, including the high
cost of living, income inequality and poverty, non-participation, deficient social
services and infrastructure, and environmental pollution. Remarkably, the weight
of economic factors was lower for migrants born in the North, who emphasized
the frustration of living there, discomfort, and family circumstances (Vlasova
and Petrov 2010, 168, 179).

The Statistical Bureau of Zabaykal'skii krai registered the following push factors
for out-migration (30,688 people in total) in the region in 2012: (1) personal reasons
(13,906), (2) employment (7,941 people), (3) education (5,246 people), (4) return to
a former place of residence (1,563). Personal reasons (primarily family issues) were
also the main push factor for the departure of interregional and interdistrict migrants.
Potential migrants in Tynda also mentioned unemployment, low salaries and living
standards, and unfavorable climate as the main causes of the current migration loss
in the BAM region.

The remaining *bamovtsy* and other populations stayed in the region for different
reasons. While some of them did not have housing in their home regions, others
have developed attachments to their local communities, family and friendship ties,
or stay for their "love for local nature." The social support provided to the residents

of the BAM region include a Northern wage premium (*severnaya nadbavka*), compensation of travel expenses connected with medical treatment, education, and recreation (the latter on a biannual basis), and other minor benefits usually provided to the residents of the Far North and the territories with similar climatic conditions. There are also several state programs in the region targeted at the socioeconomic development of indigenous people, relocating the BAM population to new housing, and relocating of the population from the North to "climatically favorable zones." However, these projects have recently been cut back to such an extent that, in practice, only a few families in Tyndinskii rayon, which has a total population of over 15,000 people, get new apartments or relocation subsidies each year.

> There is still a considerable outflow of population. State support programs don't work in the Far East. The average salary is 38,000 rubles and the prices are definitely high. That is why people leave and look for jobs. The [climate] conditions here are also harsh: we start wearing winter clothes in October. Only those remain here who don't have anywhere to leave for or whose heart is chained.[21]

Currently, ethnic Russians are the dominant group in BAM communities, with a rather insignificant proportion of indigenous Evenki (5.6 percent) and other ethnic groups. In 2012, 509 Evenki, 95 Uzbeks, 19 Azeris, 13 Armenians, 6 Tajiks, and 1 Kyrgyz were registered among the 9,051 residents of Kalarskii rayon.[22]

The interethnic relations in the BAM region are regulated by a presidential decree, "On Strategy of the State National Politics of the Russian Federation until 2025"[23] and regional legislation.[24] The major state subsidies allocated through seven special programs are spent on publications, mass media, and courses in ethnic (particularly Russian, Buryat, and Evenki) languages, folk groups, museum exhibitions, and cultural events. The Government of Zabaykal'skii krai has a work group focused on harmonizing interethnic and interreligious relations that monitors relevant activities of local and regional authorities, NGOs, and mass media. According to interviews with government officials working on ethnic and religious issues, the Northern districts of the BAM area have traditionally been characterized by ethnic and cultural tolerance. At the same time, the increasing inflow of migrants to the region raises concerns and demands adequate attention on behalf of authorities.[25]

The interregional Assembly of the Peoples of the Transbaikal region (*Assambleya narodov Zabaykal'ya*), which unites most of the registered "ethnic" NGOs, facilitates a public dialogue among the authorities, religious and ethnic leaders, academic communities, and other stakeholders. Whereas the regional cultures of indigenous peoples, Russian Old Believers,[26] and Cossacks have traditionally drawn public attention, the inflow of migrants from the former Soviet space have only recently boosted the emergence of their diasporas and NGOs.[27] The most prominent "ethnic" organization registered in Kalarskii rayon is a local branch of the Russian Association of Indigenous Numerically Small Peoples of the North and the Far East (RAIPON), which has been protecting Evenki rights since the 1990s.

A regional NGO called the Union of BAM Veterans, headed by a local journalist and a museum specialist, commemorates the history of the construction of the BAM and represents the interests of local *bamovtsy*. Other than a few groups in Chita, there are no organizations promoting the cultures and interests of the long-distance labor migrants working in the BAM region.

Authorities both in Novaya Chara and Tynda emphasize the ethnic and cultural diversity and tolerance, high educational level, and social cohesion among the local population. In interviews, they also describe qualities perceived as characteristics for each ethnic group in their districts.

> Interethnic or interreligious conflicts are nonsense for our district. The ethnic diversity of the district was facilitated by the BAM construction. Evenki have always been hospitable and nice. … They will welcome you in taiga with some tea. Our *gurany* are stubborn and complaining as usual. But the major population is those who came during the BAM construction – Komsomol members, volunteers, sometimes reckless adventurists. They came, married, and had their children born here. Each station was built by its own republic. … We don't have any kind of those conflicts that the TV shows in the Caucasus, although the regional administration is very concerned about interethnic relations. We all have been living together – we came when we were 20–22, we slept together and ate from the same plate.[28]

> Tyndinskii rayon stands out for its diverse population – there are lots of people with different professional and ethnic backgrounds from everywhere here. Despite the fact that we are considered a far periphery, children get a quality education here, because teachers came from across the (Soviet) Union, highly qualified education specialists. The same is true to engineers. Tyndinskii rayon is lucky because its people have good expertise.[29]

As indicated above, the final segment of BAM builders consider themselves to be local. When recalling BAM history, they refer to themselves as *bamovtsy*. However, in many other contexts they stress their belonging to local communities by ascribing to themselves the identities of *aborigeny* or *gurany*.

Evenki living in the BAM region have historically been considered the most tolerant, peaceful, and complacent population in terms of interethnic and interreligious relations. They appreciate the cultural and ethnic diversity of BAM communities, propagating the principles of sharing and "living like one family." The involvement of Evenki in the international indigenous rights movement and the rise of their self-consciousness in the 1990–2000s led some indigenous leaders to voice concerns regarding social discrimination, land rights, and environmental issues. In Kalarskii rayon, Evenki are worried about industrial encroachment on their lands and claim discrimination in job opportunities and education. Evenki narratives about the BAM, *bamovtsy*, and migrants typically include motifs related to environmental pollution and destruction.

They are all temporary residents (*vremenshchiki*). I have seen how they destroy the nature in Chara. They live for the day. ... They still have these habits: they fish with nets and collect berries with scrapers taking more than they can carry and preserve.[30]

In other cases, Evenki residents of BAM towns blame non-indigenous teachers for creating and circulating negative stereotypes of Evenki reindeer herders, which harm the social prestige of traditional activities and undermine the ethnic pride of Evenki children. Evenki social status seems higher in Tyndinskii rayon, which has a more sizeable and well-represented Evenki population, visible not only in reindeer herding, but also in administration, education, culture, and other spheres. There are several large-scale events regularly conducted in the district to promote Evenki culture and language. The district authorities also seem to pay more attention to traditional activities, ethnic tourism, and indigenous rights, which helps to mitigate emerging social tensions.

A certain degree of xenophobia among local residents in relation to *priezzhie*, shift workers, and labor migrants from Russia and the post-Soviet space is also connected with ecological concerns and competition for jobs. In Kalarskii rayon, local residents are concerned about potential negative impacts of developing mining and transportation infrastructure for the Udokan copper mine, fearing that it would lead to increased alcoholism, drug use, and crime in their area.[31] The competition for jobs is connected with the fact that migrants tend to be more successful in terms of employment, especially in the spheres of trade, services, and extractive industries. While the companies and authorities, apparently, lack feasible strategies for increasing employment of the local population and mitigating potential social risks presented by industrial projects, such social concerns and tensions prevail.

The BAM curse or social sustainability?

The concept of a resource curse, widely used in social sciences, rests on the argument that the countries endowed with great natural wealth tend to lag behind comparable countries in terms of long-run GDP growth and other indicators of socioeconomic development (Tompson 2006, 189). Extraction of natural resources is by definition unsustainable due to their quick depletion. Therefore, as a country runs down the available "natural capital," it has to invest in other types of capital, particularly social and human resources, in order to move to a path of sustainability (Kronnenberg 2004, 405). Today, the resource curse is observed in many countries and communities dependent on resource extraction, with Russia's Northern BAM region providing an excellent case study for this phenomenon.

State socioeconomic development strategies for Siberia, the Baikal region, and the Russian Far East[32] are targeted at stabilizing the current population levels, developing the community, supporting indigenous peoples, and diversifying local economies. At the same time, they foresee increasing extraction of mineral resources alongside industrial, technological, and infrastructural development of the Northern territories. The expected growth of the transportation infrastructure and the (re)construction of the BAM second track presume increased exports of raw materials to Asian markets. In the future, the BAM complex, with its network of side-tracks and roads, should ensure access to gas fields and metal deposits in the region. Thus, the state strategies pursue development paths leading in opposite directions. In practice, most development programs focus on exploiting mineral wealth and underestimate the value of social capital. While developing technologies and infrastructure for extractive industries, lawmakers, authorities, and companies slash investments in human resources – education, social infrastructure, services, and other important aspects of community development.

The social sustainability agenda used in political and academic discourses includes a commitment to enhance education and provide the new skills required for the "knowledge-intensive" economy, revamping employment policy to create "more and better jobs," modernizing social protection to accommodate challenges faced by welfare states, fighting poverty, and promoting equality and social inclusion (Colantonio 2007, 6). While adopting some of the clichés and buzzwords from the global mainstream political debates on sustainability, Russian development strategies and their implementation mechanisms overemphasize economic efficiency at the expense of social issues. The sporadic use of the terms "sustainable development" and "social sustainability" by authorities and company leaders in the BAM region does not mean that any tools are included to measure the sustainability of current programs and policies.

BAM construction and operation have had diverse social and ecological impacts on local communities. The prospecting stage preceding construction demonstrated the best practices of involving the local, particularly the indigenous, population in industrial projects throughout the whole history of the BAM region. In Kalarskii rayon, Evenki reindeer herders worked as porters of food, goods, and soil and stone samples for geological and engineering organizations.[33] At later stages, however, construction was carried out by a migrant labor force, with minimal participation by the local population.

However, the BAM has become an indispensable means of communication between indigenous villages, administrative centers, and reindeer herders' camps. Its social and transportation infrastructure has become invisibly, but deeply, integrated into the everyday life practices and transportation schemes of the local population. After the demise of the local aviation sector and degradation of other transportation routes in the North in the 1990s, settlements lying along the BAM or connected to it by permanent roads were left in the best position. Remote taiga areas, however, experienced increased poaching, especially among the non-indigenous population, which also increased social tensions between *aborigeny* and *bamovtsy*. Since the launch of the railroad, its growing infrastructure has also had a negative impact on the taiga landscape and Evenki traditional land use. The BAM polluted the fragile Northern taiga, interrupted animal migration routes, and damaged reindeer pastures and hunting grounds (Anderson 1991; Fondahl 1998).

Future development plans, dating back to the Soviet era, for the BAM region include construction of the second track and multiple side-tracks of the railroad, as well as the establishment of industrial clusters. One projected cluster will be located near the Udokan deposit in Northern Kalarskii rayon and includes a large processing plant and a network of the largest mineral deposits found in the region. This large-scale industrial project foresees (re)construction of the airport, roads, and social infrastructure and creating over 20,000 jobs, most of which, however, are to be occupied by outside shift workers (Polyakov 2013). This ambitious project will need significant state and private investments, which, as of late 2015, have not been secured.

BGK and SUEK, the two biggest companies presently operating in Kalarskii rayon, employ shift workers from other parts of the region, plus a few permanent residents from Udokan village, located by the mine and the administrative center of Novaya Chara. In Tyndinskii rayon, the mining industry employed 3,094 local residents, making it the largest employer, followed by railroad construction and maintenance, which employed 1,074 residents in 2012.[34] Remarkably, mining companies very rarely recruit their labor force from the indigenous population. While local authorities argue that indigenous and local residents lack the requisite education and professional skills, some company leaders propagate negative stereotypes about indigenous people.[35] In turn, Evenki NGOs blame the companies for discrimination and indifference toward their traditional culture and economic activities, as well as to the pollution and devastation of natural resources.

While extractive companies in different districts provide different employment opportunities for the local population, their overall contribution to community development is limited. Usually they act as sponsors of small-scale social projects and regular events, such as the Reindeer Herders' Day celebrated in indigenous villages, sports competitions, tourist festivals, construction of sports and cultural centers, and renovation of housing and administration buildings. Priisk Solov'evskii in Tyndinskii rayon has demonstrated the best development practices by reconstructing the social infrastructure of the rural community of Solovievsk. Authorities in Kalarskii rayon have pinned their hopes for new roads and housing on BGK, a growing company that has not yet provided any significant support to the local communities.

"Traditional industries," including reindeer herding, hunting, and associated activities, as well as the emerging field of ethnic tourism, also offer prospects for economic diversification and social sustainability of BAM towns. The authorities and heads of indigenous enterprises in Kalarskii rayon have been discussing the establishment of a reindeer breeding herd and a herder's union that would consolidate small-scale indigenous *obshchinas* and qualify for a significant state subsidy. The production of reindeer meat and skins for souvenirs and clothing, as well as the procurement of deer antlers for the pharmaceutical industry, could also be profitable and provide additional jobs and social security for herders in future.

"White Deer" is another project coordinated by the agro- and ethno-tourism center at the secondary school in Kuanda, Kalarskii rayon.[36] In addition to educational programs in traditional industries, the center organizes tours to reindeer herders' camp in the taiga and sponsors museum exhibitions on Evenki culture.[37] In Tyndinskii rayon, tourists can visit the "Evenki Village" museum that opened in the Pervomaiskoe indigenous settlement in March 2012. The site includes an Evenki cultural center, a souvenir shop, and exhibitions of traditional and contemporary Evenki dwellings and artifacts.[38] Currently, these initiatives are supported by the socioeconomic development programs for indigenous peoples of the North and employ some local activists and educators. However, more significant state and private investments could further promote ethnic tourism, increase employment, and draw public attention back to indigenous issues.

The developing mining industry has increased competition for natural resources, especially land. The growing industrial infrastructure encroaches upon traditional lands of indigenous *obshchinas* involved in reindeer herding and hunting. In Kalarskii rayon, the current policies, focused on resource extraction, are implemented by a coalition of local authorities and companies. Indigenous NGOs may assert land claims at public hearings, but they tend to be disregarded by company leaders. In terms of rival land claims, the local authorities give unconditional preference to large industrial companies. They foresee the relocation of Evenki reindeer herding enterprises and the alienation of their traditionally occupied lands for industrial development, without compensation. Such encroachment of expanding industrial infrastructure on traditional lands reduces opportunities for the development of large-scale reindeer herding and ethnic tourism, while increasing social tensions, especially between indigenous and non-indigenous populations.

Conclusions

The BAM region represents a remarkable site for an anthropological inquiry into the Soviet and post-Soviet industrialization history of the Russian North. Contemporary BAM settlements, with their ethnically and culturally diverse population, resemble typical single-industry towns dependent upon the functioning of the railroad and mining industries. Indigenous people, BAM builders, and newcomers are three distinct groups forming the social networks of BAM communities.

The original ethnic fabric of the region was woven of the indigenous population – Evenki and other Tungus-speaking minorities and Russian Old Settlers. Their subsistence activities – hunting, reindeer herding and cattle-breeding – still play an important economic and cultural role. Presently, there are dozens of small-scale indigenous enterprises leading semi-nomadic lifestyles on their traditional lands, which are often encroached upon by extractive industries and infrastructural developments. However, labor migrants have also played a major role in the formation of the region's cultural tapestry.

The construction of the BAM settlements and infrastructure triggered the most significant population influx in the region's history, drawing potential workers from across the former USSR. The BAM has considerably shaped local identities and communities, and BAM builders and their descendants now constitute the majority of the population in the settlements along the railroad, as well as a visible proportion of residents in adjacent mixed communities. Today, a second wave of migrant workers from the post-Soviet space has flooded the region to revive and expand the railroad, housing, industrial plants, and other infrastructure. In many cases, they experience precarious working conditions and confinement to specific economic sectors. Another set of new incomers are long-distance commuters (see Chapter 5): the extractive industries operating in the region increasingly employ qualified workers from all over Russia, using the fly-in/fly-out shift work system as their main method of operation.

Aborigeny, bamovtsy, priezzhie, and other names serve as identity markers, ascribed and self-ascribed in the course of identity politics and intergroup relations. Soviet nation-building policy has proven effective in forging the social coherence of a multiethnic labor force using the construction of the BAM as an experimental social engineering project. In this process, ethnicity has served as an important factor in forging the culturally diverse, yet integrated local communities with histories of long-term, peaceful co-existence and cooperation. The present cultural and religious tolerance still carries significant potential for social sustainability of the BAM region. While the boundaries between "locals" and "migrants" are flexible and penetrable, having ties to the local community is an asset, which is especially valued in interrelations between the main groups and stakeholders.

The 2012 national political strategy and recent regional development programs have yet to be tested by new socioeconomic reality and trends. Local authorities and mining companies operating in the region pursue resource-dependent economic policies, often neglecting social and human capital as the most important factor for sustainable community development. Such unwise policy leads to mutual negative stereotypes and social tensions between local and non-local residents, based on competition for jobs, lands, and state support. Official support of traditional economic activities and ethnic tourism, recognition of indigenous rights, and creation of local mechanisms for integration of new migrants would create more equal employment opportunities and access to land and other resources and redirect BAM communities to a path of a more sustainable socioeconomic development.

Notes

1 This chapter is based on the field and archival study conducted in the city of Chita, Kalarskii rayon, Zabaykal'skii krai, and the city of Tynda in Amurskaya oblast in September–October 2013. The preliminary results of the study were presented and discussed at the Arctic Frontiers Conference in Tromso, Norway, on January 22, 2013. The author thanks the George Washington University for supporting the research upon which this chapter is based. During the final stages of preparing this publication, support from the Austrian Science Fund (FWF, project P27625-G22, "Configurations of Remoteness (CoRe): Entanglements of Humans and Transportation Infrastructure in the Baykal–Amur Mainline (BAM) Region") has been important.

2 Vserossiiskaia perepis' naseleniia 2010 (accessed October 10, 2014). www.gks.ru/free_doc/new_site/perepis2010/croc/perepis_itogi1612.htm.
3 Administration of Tyndinskii rayon, 2013. *Itogi sotsial'no-ekonomicheskogo razvitiia Tyndinskogo raiona za 2010–2012 gg.* (Hereafter, *Itogi.*)
4 Vserossiiskaia perepis' naseleniia 2010.
5 Statistical Bureau of Amurskaya oblast. 2013. *Sotsial'no-ekonomicheskoe polozhenie gorodskogo okruga Tynda za ianvar'–dekabr' 2012. Doklad.* (Hereafter *Sotsial'no-ekonomicheskoe polozhenie.*)
6 Expert Institute. 1999. *Monoprofil'nye goroda i gradoobrazuiushchie predpriiatiia.* Moscow, 12.
7 *Itogi*, 62.
8 Vserossiiskaia perepis' naseleniia 2010.
9 Administration of Kalarskii rayon. 2013. "Spisok lits korennoi natsional'nosti, prozhivaiushchikh na territorii Kalarskogo Raiona."
10 *Itogi*, 17.
11 Department of Agriculture. Administration of Tyndinskii rayon. 2013. *Dannye po rodovym obshchinam Tyndinskogo raiona*; Tynda Administration, Department of Agriculture. 2013. *Pogolov'e olenei po Tyndinskomu raionu.*
12 Zabaikal'skii krai. Ministry of Territorial Development. 2013. *Informatsiia na soveshchanie po problemam korennykh malochislennykh narodov Severa v Zabaikal'skom Krae na temu "Gosudarstvennaia podderzhka korennykh malochislennykh narodov Severa v Zabaikal'skom Krae,"* 2.
13 Author's interview with respondent L.A.S., 2013.
14 *Migratsiia naseleniia Zabaikal'skogo kraia: statisticheskii sbornik.* 2013. Chita: Zabaykalkraystat. (Hereafter, *Migratsiya.*)
15 Author's interview with respondent V.V.A., 2013.
16 *Migratsiia*, 11–13, 32.
17 *Itogi*, 17.
18 *Sotsial'no-ekonomicheskoe polozhenie*, 31.
19 *Migratsiia*, 33.
20 Zabaikal'skii krai, Department of Federal Migration Service. 2013. *Svedeniia po prebyvaniiu inostrannykh grazhdan iz stran blizhnego zarubezh'ia na territorii Zabaikal'skogo Kraya (s tsel'iu "rabota").*
21 Author's interview with respondent L.A.S.
22 Administration of Kalarskii rayon, Decree #107. *Ob utverzhdenii munitsipal'noi tselevoi programmy "Ekonomicheskoe i sotsial'noe raztivie korennykh malochislennykh narodov Severa (2013–2015)."* Ratified March 15, 2012, 4–5; Zabaikal'skii krai, Governor's Internal Administration, Department on Relations with Social, Religious and Public Organizations. 2013. *Natsional'nosti v munitsipal'nykh raionakh i gorodskikh okrugakh.*
23 Decree of the President of the Russian Federation, "On Strategy of the State National Politics of the Russian Federation until 2025," N 1666, ratified on December 19, 2012.
24 For example, Law of Zabaykal'skii krai, "On State Support of Socially Oriented NGOs in Zabaikal'skii Krai"; Governor's Decree, "On Monitoring of Interethnic and Interreligious Relations and Responding to Manifestations of Religious and Ethnic Extremism"; Governmental Act, "Action Plan for Harmonization of Interethnic Relations in Zabaikal'skii Krai in 2011–2013."
25 Author's interview with respondent M.S.K., 2013.
26 A Russian religious minority including followers of traditional ritual practices, which emerged as a group after the Orthodox Church Reform conducted in 1653–1656.
27 Zabaikal'skii krai, Governor's Internal Administration, Department on Relations with Social, Religious, and Public Organizations. 2013. *Obespechenie administrativno-pravovogo regulirovaniia sostoianiia mezhnatsional'nykh otnoshenii na territorii Zabaikal'skogo Kraia*, 1–3.

28 Author's interview with the head of the Kalarskii rayon Administration.

29 Author's interview with an official in the Tyndinskii rayon Administration, a former BAM builder.

30 Author's interview with respondent N.P.G., 2013.

31 SRK Consulting. "Vzaimodeistvie s zainteresovannymi storonami v ramkakh proekta razrabotki Udokanskogo mestorozhdeniia medi: otchet o poseshchenii raiona proekta," www.bgk-udokan.ru/upload/doc/Udokan_SEP_RU_Version_3.pdf, accessed on December 19, 2015.

32 "Strategy for Socioeconomic Development of Siberia until 2020"; "Strategy for Socioeconomic Development of Baikal Region and the Far East until 2025."

33 Author's interview with L.V.M.

34 *Itogi*, 19–20.

35 Author's interview with respondent G.M.Z., 2013.

36 Belyi Olen'. "Tsentr agroetnoturizma" (accessed October 10, 2014). http://whitedeer .ru/index.html.

37 Administration of Kalarskii rayon, Decree #107. *Ob utverzhdenii*, 7.

38 *Itogi*, 31.

8 Trajectory of a city

Murmansk and its transforming diversity

*Marlene Laruelle, Sophie Hohmann,
and Alexandra Burtseva*

Murmansk is unique among Russia's Far North cities. It is the largest Arctic city in terms of population (300,000 inhabitants), an urban engineering feat that stretches for more than 20 kilometers along the rocky coast of the Kola Bay at an extreme latitude (68° N). Yet, for an Arctic city, Murmansk benefits from a relatively friendly environment: just 50 kilometers from the Barents Sea, the city hosts the only Arctic warm-water port thanks to the Gulf Stream. Even if life is regulated by polar nights and days, Murmansk's mild winters and its relative proximity to central Russia make it a unique place on Russia's Arctic coasts, less remote than other big cities such as Norilsk or Yakutsk.

The city offers a complex, multifaceted society, with an urban identity constructed entirely during Soviet times but not linked to an extractive industry. This relatively diversified social and economic fabric can only be compared, throughout Russia's Far North, with Yakutsk. In this chapter we explore Murmansk's population mobility patterns in the post-Soviet era, and we put this mobility in the context of the city's character. First, we look at the Soviet-era waves of population movements then, second, at the current socioeconomic structure of the city and the region. In the third part we explore new post-Soviet migration flows, particularly the arrival of migrants from the South Caucasus and Central Asia, and the progressive emergence of a diaspora identity that is well integrated into Murmansk's own identity.

History of population movements on the Kola Peninsula

The Saami were the only people to live on the Kola Peninsula for some 10,000 years. However, traces of scattered Slavic enclaves have been confirmed by written sources and archeological digs. Under Ivan the Terrible, the Muscovy principality tried to consolidate its hold on the territories of the north and proceeded with what the historian Pavel Fedorov called "a monastic colonization" (2009, 64). The southern portion of the peninsula is indeed part of the "Pomor" world, a name attributed to Russian populations settled around the perimeter of the White Sea from the sixteenth century on, and who lived mainly on fishing, agriculture, hunting, and the sale of furs – livelihoods still pursued in the small towns of Kola and Kandalashka. In later centuries, with the opening of the Baltic by Peter the Great and the competition with Sweden, Russia sought to consolidate its power over the Murman region – taken from the Russian name for the

Scandinavians, *Normand* – and to acquire open ocean ports. In the nineteenth century, Komis moved into the region, as did some Nenets fleeing northward from the White Sea region.

The first signs of urbanization in the Kola Peninsula began in the last years of the Tsarist era. In the 1890s Minister of Finance Sergei Witte, presiding over the Russian Empire's extensive industrialization, called for the construction of a port on the Barents Sea, but only with the onset of World War I did the Tsarist authorities act on this. Construction of the St. Petersburg–Murmansk railway began in 1915, and the small terminal fishing port of Romanov-na-Murmane was dedicated in 1916, making it the last town founded by the Tsarist regime (Mikhailov 2013). The Provisional Government renamed the village Murmansk in 1917. Murmansk and its region are therefore a pure product of Soviet industrialization: the town went from 2,500 inhabitants before the October Revolution to 120,000 just before World War II, then to 472,274 in 1989, at the last Soviet census.

Murmansk's strategic location made it an outpost of socialism that the Bolshevik regime had to defend at all costs during the Civil War against the White Army of Admiral Alexander Kolchak, which was backed by the British and the Norwegians. From 1922, port activities developed and the first fish industries emerged. During the 1920s, the authorities hoped the region would populate naturally, designated by the term of "Canadization" (*kanadizatsiya*), to refer to Canada's non-state-directed population movements (Fedorov 2009, 208 ff.). But this policy failed. Managers from the railroads, related industries, and local companies complained about constant labor shortages and the high population turnover. At the end of the 1920s, the USSR entered into a phase of rapid industrialization and adopted a system of five-year plans, a period during which the Kola Peninsula specialized in the ore industry: the geological finds on the Khibiny Mountains saw the first ores extracted at Apatity in 1931. The great metallurgic complex at Severonikel was built in 1935.

Moscow abandoned Canadization in the late 1930s in favor of a voluntarist strategy of populating the region. Unlike many other Arctic areas of the Soviet Union, prison labor (deported individuals or *spetspereselentsy* and GULAG prisoners) only constituted about 20 percent of the workforce in the 1930s–1940s (Fedorov 2009, 254–255; Mikhailov 2004, 90–91). Instead, the profound social upheavals of the period, particularly the massive urbanization and the destruction of the peasant world in central Russia, facilitated immigration to the Kola Peninsula, especially as the first financial incentives (*severnye l'goty*) were set up in 1932 to stabilize the population.[1] The city of Murmansk became the home of the Northern Fleet, as well as one pillar of *Glavsevmorput*, the central Soviet administration in charge of developing the Northern Sea Route along the Arctic Ocean. Thanks to these multiple development projects, Murmansk underwent an unprecedented demographic boom: the town went from having 11,400 inhabitants in 1928 to 117,000 in 1939 (Fedorov 2009, 259).

The destruction wrought by World War II was immense and set back the region's development. The city's memory is still heavily shaped by WWII, as seen with the Alyosha monument dominating the city landscape (Figure 8.1). Only Stalingrad was bombed more times by Germany, and in 1945 Murmansk was included on the list of 15 Soviet cities earmarked for urgent reconstruction. It returned to its prewar population and infrastructure levels in the 1950s, due in part to the labor

Figure 8.1 Alyosha, the monument to Soviet soldiers, sailors, and airmen of World
War II, which dominates the city (photo by Sophie Hohmann).

power of the GULAG prisoners, the *spetspereselensy*, and German prisoners of
war. Forced work came to an end in 1953 with the death of Stalin, but the prisoners
continued to work at the region's hydroelectric stations into the 1960s (Fedorov
2009, 349). New residents arrived thereafter only on a voluntary basis. The urban

structures grew, and many surrounding settlements were integrated into the city or better connected to it (Fedorov 2014). The three subsequent decades, from 1955 to 1985, constituted the golden age of Murmansk and the surrounding region.

The peninsula became one of the main Soviet mining centers, extracting coal, iron, and non-ferrous minerals. In 1958, 80 percent of all the phosphates used in the USSR were produced in Apatity (Fedorov 2009, 341). With the exception of Murmansk itself, nearly all the towns of the peninsula were directly connected to the mining and metallurgy industries, and the first indicators of environmental damage were visible at Monchegorsk and Nikel as early as the 1960s. Murmansk became the premier fishing port in the entire Soviet Union: its factories handled 224,000 tons of fish in 1950 and 1.2 million tons in 1980 (Fedorov 2014, 343). The militarization of the peninsula also changed the urban fabric: a northwestern outpost in the USSR face-off with NATO, Murmansk benefitted from the complete attention of the Defense Ministry and the military-industrial complex, hosting the Northern Fleet and its nuclear icebreakers, as well as significant air forces. The military training bases and airfields were strewn throughout the peninsula, with more than a dozen closed cities – Severomorsk being the largest – where military personal and their families were housed.

During these three decades, the Murmansk region saw the most significant population increase of the entire Soviet Far North. Many Komsomol (Communist Youth League) members who were sent there for work for a few years decided to stay and take advantage of the financial incentives handed out by the Soviet state to workers in the Far North (*severnye nadbavki*). The industrialization of the Murmansk region brought a massive influx of Russians from across central Russia, as well as of Ukrainians and Byelorussians, who today comprise the two largest ethnic groups (*natsional'nosti*) after Russians.[2] All found a professional niche in the local mining and fishing industries or in the military. Ukrainians were particularly attracted to the region, as their experience working in the mines of Donbass and in the marine market on the Black Sea easily blended into the needs of Murmansk and its surrounding region.

The rate of population increase in the Murmansk region was far higher than the average of the Russian Federation (RSFSR). In the 1950 and 1960s, the natural rise was 24,000 persons per year, of which around 8,000 were new arrivals from other regions. The outside arrivals slowed in the 1970s and even more sharply in the 1980s, but the population continued to increase through its own reproduction, with a birth–death ratio that was still largely positive. Just before the collapse of the Soviet Union, the region reached its maximum with 1.16 million persons in the 1989 census, including half a million in Murmansk (see Table 8.1).

Between the end of World War II and the fall of the USSR, the city's real estate market grew seventeen-fold (Fedorov 2009). This massive and rapid urbanization gave the city a uniform style, which included buildings in the Stalinist style in the city center, some wooden collective houses on the peripheries, and some rare districts of *khrushchevki* (Khrushchev-era apartment blocs). About two-thirds of the urban landscape were drab buildings from the 1970s that were typified by long rows of shoddy, concrete, mid-rise buildings without any particular color (*panel'nye doma*), and that deteriorated rapidly.

Table 8.1 Demographic evolution of Murmansk region and Murmansk city, 1926–2010

	1926	1959	1989	2002	2010
Murmansk region	23,006	567,972	1,164,586	892,534	795,409
Murmansk city	8,777	221,874	472,274	336,137	307,257

Source: P.F. Fedorov, *Kul'turnye landshafty kol'skogo severa v usloviakh urbanizatsii (1931–1991 gg)*. Murmansk: Murmansk State Pedagogical University, 2014.

As with all the other regions of the Far North, Murmansk was badly affected by the collapse of the Soviet Union. The disappearance of the planned economy, the downsizing of the Northern Benefits scheme, the dearth of jobs, dim prospects for young people, exorbitant prices for basic goods, chronic shortages of heat, gas, and electricity, and declining transportation links with the rest of the country pushed millions of Russians to leave Far Northern regions for central Russia (Heleniak 2009a, 2010). However, compared with Chukotka or Sakha-Yakutiya, Murmansk was not as severely affected. While some medium-size agribusiness complexes shrank, the main industrial complexes were able to continue functioning, even if with difficulties.

The Murmansk region lost 16 percent of its inhabitants – more than 180,000 persons – in the 1990s. That works out to an average rate of 20,000 persons per year; however, there was a record departure of 31,000 in 1992. From the start of the 2000s, the population decline slowed to around 8,000 departures per year (Bardileva and Portsel 2014, 99). In total, between 1991 and 2012 the region lost one-third of its inhabitants. This situation brought widespread transformations to both rural life – one village out of eight is now devoid of inhabitants – and cities. The social fabric was ripped, while family and professional ties were transformed. As Figure 8.2 shows, life expectancies have risen more slowly than in the rest of Russia (63 years for men, 74 for women).[3]

Socioeconomic diversity

The Murmansk region is more socioeconomically diverse than other parts of the Far North. Mining and metal processing produce 35 percent of regional revenues, which is a high figure, but it means two-thirds of the GDP originates from other sectors (Tsukerman and Goriarchevskaia 2014).

Although its population has decreased, Murmansk has avoided the perforation of the urban landscape that many of its neighbors have suffered. None of its districts have been abandoned en masse. Apart from the key sectors listed below, the city has been able to preserve some agribusiness-related industries, as well as transport and logistics companies that take advantage of the port and the railway. Its status as a regional administrative center secures state administration jobs and a developed retail sector.[4] The presence of a large contingent of military personnel, who change their placements every five years on average, also guarantees a dynamic real estate market. The neighboring city of Severomorsk, less than 20 kilometers away, with more than 50,000 inhabitants, functions almost as a

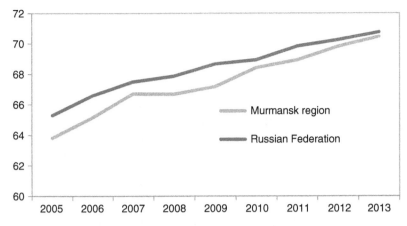

Figure 8.2 Trends in life expectancy at birth in the Murmansk region and the Russian Federation.

Source: Socio-economic indicators for all the Russian regions, 2015 (www.gks.ru/bgd/regl/b15_14p/Main.htm).

military "suburb" for Murmansk. In total, about 130,000 persons live in military towns – Closed Territorial Entities – on the peninsula (Revich et al. 2014, 202).

Because of its diversified economy, Murmansk offers a varied employment market. The service sector – such as retail and wholesale trade, motor vehicle repair, entertainment and leisure activities – was the first to adapt to the post-Soviet economic changes (Revich et al. 2014). However, there has been little progress in technological innovation, and the city's student population continues to diminish, lured away by the universities of central Russia.[5] The failure to develop the gigantic Shtokman gas deposit, the result of disagreements between Gazprom and Statoil,[6] has cut short promises of major energy developments in the Barents Sea, which would have created a new economic sector for the city.

The Northern Fleet employs around 30,000 people and has entered a modernization phase after the stagnant 1990s. Moscow's largest naval fleet, it includes most of Russia's missile-carrying strategic submarines, the largest number of icebreakers and nuclear submarines, as well as about two-thirds of the Russian Navy's nuclear force. Besides reflecting the rise of Russia's naval power, the Northern Fleet and the region's new motorized rifle Arctic brigade are responsible for protecting the country's considerable economic interests in the Arctic region, including safeguarding oil and gas installations and tanker traffic from possible technical incidents and less probable terrorist attacks (see Laruelle 2014). The Northern Fleet, FSB border guards, and Arctic patrols are therefore likely to remain one of the main sources of employment on the peninsula and are celebrated in the urban landscape (see Figure 8.3).

After the service sector and the Fleet, the third niche is the ocean itself. The Barents Sea is rich in fish resources. Russian fishing fleets, especially the Murmansk Trawler Fleet, were burgeoning in the 1970s, but today urgently need an overhaul. In the 1990s, state investment in the fisheries collapsed, exacting

Figure 8.3 Brezhnev-era apartment blocks celebrating the Soviet icebreaker Arktika and
its polar expedition of 1977.

a heavy toll. The size of the Russian fleet plummeted by half, and two-thirds
of fishing vessels still in operation no longer conform to safety standards and
have exceeded their legal lifespan. But Murmansk finds itself in a better situation
than some other big fishing ports such as Petropavlovsk-Kamchatsky. The powerful Union of Fish Producers (*Soyuz rybopromyshlennikov*) was able to invest in
new equipment, and Murmansk's ships now have onboard facilities for processing caught fish. Russian–Norwegian cooperation has been successfully manag-
ing Atlantic cod stocks and Norwegian spring-spawning herring for several years
(Laruelle 2014).

The commercial port at Murmansk (*Murmanskoe morskoe parokhodstvo*) is
dominated by the Murmansk Shipping Company, a leader in oil transportation and
transshipment in the Arctic thanks to its partial purchase by Lukoil. The company's
fleet is designed to maintain continuous, year-round exports from the company's oil
production facilities in the Timan-Pechora district. Today Lukoil is the main operator
in the Arctic Basin, with around 200 vessels of different types. The Prirazlomnoye oil
field in the Pechora Sea, the first commercial offshore oil development in the Arctic,
which began extractions in 2014 after more than a decade delay, should acceler-
ate oil traffic in the Barents Sea and benefit Murmansk (Petrova 2014). The city
also hosts institutions and administrative entities linked to the Northern Sea Route,
whose activities are on the rise, from four vessels in 2010 transporting 111,000 tons
to 129 for 2014, transporting 1.6 million tons.[7] The traffic is mainly confined to the
western section of the Russian Arctic coast, between Murmansk and Dudinka, on
account of oil-related activities but also rising exports of minerals, roundwood, lum-
ber, pulp, and paper. With the increase in gas production, the Barents Sea is bound

to become the most dynamic part of the Russian Arctic and the most congested with ships and vessels. International shipping remains limited, but domestic shipping is rapidly increasing alongside the many energy-related projects that Moscow is developing in the Arctic and sub-Arctic regions. The city also hosts geology services and exploration outfits for the continental shelf such as Arktikmorneftegazrazvedka and Morgeo, as well as the two agencies responsible for cleaning up after oil and gas spills, Murmansk Basin Emergency Rescue Service and the company Ekospas-Murmansk, and two shipyards. The port remains therefore the shaping element of Murmansk's urban landscape (see Figure 8.4).

Outside of Murmansk, the Kola Peninsula has a much less diversified economy. Eight towns are categorized as monocities (*monogorod*),[8] which totals 44 percent of all Russia's Arctic monocities. A city is considered as monocity when 25 percent of its population works in its primary (*gradoobrazuyushchii*) industry, more than 50 percent of total town production is generated by this industry, and more than 20 percent of the municipal budget is dependent on it (Didyk, Riabova, and Emel'ianova 2014, 11–12). These monocities cannot function in a free market economy without state intervention: if the enterprise goes bankrupt, the town collapses along with it. About 20 percent of the peninsula's population lives in these monocities. More than 70,000 persons have moved away since 1989, similar to the rate of departure of other cities of the region (Didyk, Riabova, and Emel'ianova 2014).

Three monocities – Monchegorsk, Nikel, and Zapolyarnyi – are part of the Kola Mining and Metallurgical Company, which was absorbed by Norilsk Nickel in 1989, just prior to the fall of the USSR. Norilsk Nickel is thus the largest employer on the peninsula, with around 13,000 direct employees.[9] In Kovdor, Kovdor Ore-Mining Integrated Works is part of the holding company Eurokhim,

Figure 8.4 View of Murmansk port.

which is the city's largest employer, and the second-largest extractor of apatite compounds in Russia. The Olkon mining and processing plant, part of Severstal, Russia's largest steel company, is based in Olenogorsk. Employment in the large metallurgy companies fell in the 1990s, but stabilized thereafter, and has managed to keep up its level of employment since the economic crisis of 2008.

The Murmansk region has yet to fully realize all of its natural resources potential, such as untapped rare earth minerals, especially eudiyalite, an extremely rare mineral that can be surface-mined. All monocities face the same structural issues: although the primary enterprise provides decent salaries and stable careers, the town becomes less attractive to job seekers. The social fabric is disintegrating (the peninsula's smallest monocities, Kovdor and Revda, have lost one-third and 40 percent of their population, respectively), unemployment is higher than in the rest of the region (women are especially hard hit), housing and social services are deteriorating, and the cultural and educational offerings are minimal.[10]

Some cities have escaped the curse of the monocity. Apatity, the second-largest city in the region with 59,000 inhabitants, is dominated by PhosAgro, one of the biggest producers of phosphate, extracted from the Khibiny Mountains, the largest supplier of the two most popular phosphate-based fertilizers used in Russia. But the city is also home to the Kola Research Center, which employs more than 1,500 researchers. In the south of the peninsula, Kandalaksha and Kola are two historical towns representative of the Pomori culture, founded on agribusiness, wood furniture, tourism, and, for Kandalaksha, transportation and cross-border activities with Finland.

Since 2008 the Murmansk region has been developing eco-tourism, but progress in this sector has been rather modest. The region has undeniable potential in terms of outdoor sports – skiing, kayaking, water sports, ocean diving, and fishing – particularly for salmon – but for the moment its international visibility is still limited. The Murmansk brand has still not been widely exported; it faces serious competition from Norway, which has long been the leader of "green" tourism in the Arctic region. So far Murmansk appeals only to Russian tourists or tourists from neighboring Scandinavian countries, who can take advantage of the difference in living standards to enjoy cheap vacations. In 2013, the city received 260,000 Russian visitors and 45,000 foreign visitors (Zhelnina, 2014, 42).

Plans to develop ethnic tourism are also still in the early stages. Although many of these plans have focused on the Lovozero region, where the majority of the Saami population is concentrated, they have not been profitable (Aleksandrovna and Aigina 2014). The establishment of "Saami Village" (*saamskoe selo*), a place where tourists can live like traditional reindeer herders, does not have enough facilities for large-scale tourism, and the Saami themselves want to be certain that they will not be the "folklorized" losers of ethno-tourism. The various Saami feast days, such as Feast of the North, and the Saami World Games nonetheless enable the Lovozero region to increase its visibility, mainly within the Scandinavian world, and the neighboring countries' Saami communities. In any case, ethno-tourism will not solve the main dilemma of the Saami, which is the loss of the Saami language by the young generations (Ivanishcheva 2014, 246–251) and the almost total disappearance of traditional reindeer herding.[11]

Waves of migration to Murmansk

This broad overview of Murmansk's economic situation provides the background for the region's overall demographic dynamics. The Murmansk region is the most highly urbanized zone in all of Russia, with about 93 percent of its inhabitants being city dwellers. It also has among the country's youngest population (37.6 years on average, compared to 39 for the whole of Russia),[12] even if it is in a phase of aging. The working population (68 percent) is also larger than in many other regions of the country, and the proportion of men higher (49 percent for Murmansk, 46 percent for the rest of Russia) (Fedorov 2009, 98–99). Compared with the other Arctic regions of Russia, Murmansk is privileged for its relative proximity to the population centers and wealthy areas of central Russia. It is also advantaged by its socioeconomic diversity. It is not dependent upon a single type of industry; instead, it has a diversity rarely found in the Arctic: military bases, extractive industries, transformative industries, commercial fleets, and tourism potential.

New migrant flows from southern countries

However, like many other Arctic regions, Murmansk and the Kola Peninsula face difficulties in maintaining demographic growth and in meeting the demand for labor. Young people born in the region tend to leave for more southern cities such as Petropavlovsk, in Karelia, or St. Petersburg and Moscow. According to the statistical state agency Rosstat, the gradual aging of the population and the high number of people leaving for more hospitable regions will lead to a further loss of 100,000 inhabitants between 2015 and 2020 (Revich et al. 2014, 203). In 2011, the regional authorities of the North-West federal district adopted a "Strategy of Socioeconomic Development" that highlights migration as a source of demographic growth.[13] In 2014, the Murmansk oblast administration announced more than 40,000 available jobs (up from 37,000 in 2013).[14] Statistics provided by the local branch of the Migration Federal Service confirm that every year, between 5,000 and 7,000 foreigners get either a long-term work permit (*vid na zhitel'stvo*), or a short-term registration card (RVP, *razreshenie na vremennoe prozhivanie*) to make up the labor shortfall.[15]

The traditionally high mobility of the Far North has been heightened in recent years by new migration flows from the south. These new migration patterns emerged with the collapse of the USSR. As Figure 8.5 shows, today Murmansk attracts migrants not only from Russia's other regions but also from countries of the Commonwealth of Independent States (CIS), mostly individuals of working age. Populations originating from the southern republics of the former USSR – Azerbaijan and, to a lesser extent, Central Asia – have come to the region in significant numbers looking for work, often following the first wave of migrants who settled in the region in the 1980s (Laruelle 2016). According to the 2010 census, Azerbaijanis constituted more than 40 percent of all "southerners," followed by Armenians (18 percent) and Uzbeks (12 percent).[16] Murmansk city received about 20,000 Azerbaijanis, but this number seems to be decreasing (see below), and about 8,000 Central Asians. Among the latter, the Uzbeks are by far the most

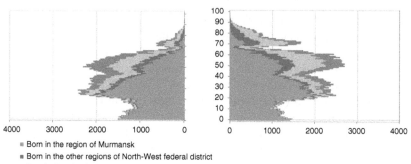

Figure 8.5 Urban population of Murmansk region based on place of birth, 2010 census.

Source: Rosstat, 2010 census.

numerous, with around 7,000 persons in Murmansk itself – the first wave were Uzbeks from the south of Kyrgyzstan, fleeing the ethnic violence of 2010. They have been followed by around 500 Tajiks and 300–500 Kyrgyz.[17]

Migrant flows are far from static. Since 2013 the flow from Azerbaijan has slowed. According to the VAK association (see below), the Azerbaijani diaspora in Murmansk has decreased from 20,000 in the early 2010s to 14,000 people in 2014–2015.[18] This decrease can be explained by several concomitant factors: the Russian economic crisis that has limited opportunities for the latest newcomers to find a sustainable professional niche, and the – very relative – well-being of some regions of Azerbaijan, which encourages potential migrants to stay at home. It is also possible that other Russian cities attract more new migrant flows from Azerbaijan than Murmansk.

However, immigrants from Central Asia are rapidly replacing the Azerbaijanis. The Russian economic crisis has affected fewer Central Asian migrants because the domestic issues in their republics of origin have not improved, and the gap between a Russian salary, even if lower, and a Tajik, Uzbek, or Kyrgyz one is still large enough to motivate departures. Table 8.2 and Figure 8.6 show this slowdown of flows from Azerbaijan and the rise of those from Central Asia. It also illustrates how attractive the city remains for Ukrainians and Byelorussians, who come mostly to join members of their family already located in the region. The inflow of Ukrainians noticeably jumped with the massive refugee flows that accompanied the 2014 Ukrainian crisis. In 2013, people getting Russian citizenship in the Murmansk region were primarily from Ukraine, Armenia, Kazakhstan, Uzbekistan, Kyrgyzstan, and Azerbaijan.[19]

Murmansk's "migrant aristocracy": the Azerbaijanis

Azerbaijani migrants come to Murmansk for one of two reasons: military service or work. For decades, Soviet citizens of all nationalities performed their military service with the legendary Northern Fleet. Among the Azerbaijani diaspora of

Table 8.2 Registered labor migrants in Murmansk city in 2011–2013 by country of origin

Country of origin	Official employment registration		
	2011	2012	2013
Azerbaijan	432	478	564
Armenia	142	139	277
Belarus	532	1721	968
Kazakhstan	14	59	100
Kyrgyzstan	492	531	902
Moldova	81	189	343
Tajikistan	313	324	884
Uzbekistan	966	1,359	1,939
Ukraine	295	579	968

Source: Rosstat, "Natsional'nyi sostav naseleniia po regionam Rossii," *Vserosiiskaia perepis 'naseleniia*, 2002 and 2010, www.gks.ru/free_doc/new_site/perepis2010/croc/Documents/Vol3/pub-03-01.pdf, www.perepis2002.ru/index.html?id=1, and "Raspredelenie naseleniia po natsional'nosti I rodnomu iazyka," 1959, 1970, 1979 and 1989 perepis', http://demoscope.ru/weekly/ssp/rus_nac_89.php?reg=4, http://demoscope.ru/weekly/ssp/rus_nac_79.php?reg=4, http://demoscope.ru/weekly/ssp/rus_nac_70 .php?reg=4, http://demoscope.ru/weekly/ssp/rus_nac_59.php?reg=47.

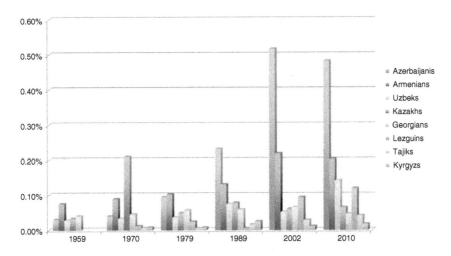

Figure 8.6 Ethnic groups from Caucasus and Central Asia in Murmansk oblast, 2010 census.

Source: 1959–1979: RGAE RF (byv. CGANH SSSR), fond 1562, opis' 336, http:// demoscope.ru/weekly/ssp/rus_nac_79.php?reg=4; 1989: Rabochii arkhiv Goskomstata Rossii. Tablitsa 9c. Raspredelenie naseleniia po natsional'nosti i rodnomu iazyku, http:// demoscope.ru/weekly/ssp/rus_nac_89.php?reg=4; 2002: Vserossiiskaia perepis' naseleniia 2002 goda. Natsional'nyi sostav naseleniia po regionam Rossii, www.perepis2002. ru/index.html?id=1; 2010: Vserossiiskaia perepis' naseleniia 2010 goda. Naselenie po vozrastnym gruppam, polu i urovnia obrazovaniia po sub"ektam Rossijskoi Federatsii.

Murmansk, a generation of newcomers, born in the 1960s, served in the Fleet in the 1980s and then settled permanently in the region.[20] Many explain their decision to remain by the sociopolitical changes of the *perestroika* years and the early of the 1990s: interethnic tensions in the Caucasus, economic hardships during the introduction of the market economy, and the opportunity to advance as a contract soldier after the required two years of service.[21]

The paths taken by Azerbaijanis who had left their republic for Russia in Soviet times plainly align with identifiable professional career tracks. In the 1950s and 1960s, Azerbaijani engineers and gas and oil specialists (*neftyanniki*) headed north. They migrated toward the large Russian petroleum deposits of western Siberia, then to the Far North, and many of them settled there permanently. They are nonetheless barely represented in Murmansk, as the region has no oil and gas industry. A second wave, which formed in the 1970s and 1980s, consisted of fruit, vegetable, and flower merchants (Yunusov 2003). This flow was more circular, allowing for more regular visits between the republic of origin and the Russian city of settlement.

With the end of the USSR, the "ethnicization" of labor continued. In Murmansk, the Azerbaijanis worked at wholesale fruit and vegetable markets (*ovoshchnye bazy*), and opened cafes and restaurants featuring their native dishes. This niche is not only ethnicized, but regionalized, for the vast majority of them stem from the remote regions of Massali and Lenkoran.[22] Already in Soviet times Azerbaijani traders in Russia originated from these southern regions, as well as from the northeast of Azerbaijan (Gusar and Khatchmas – mainly the Lezguines) thanks to their accumulated agricultural expertise. Azerbaijanis hailing from Nagorno-Karabakh, especially from Agdam, who arrived in Murmansk around the end of the 1980s and the beginning of the 1990s during the conflict with Armenia, worked in the clothing trade or at the flea markets. Through the different generations of migrants one can therefore read the history of Azerbaijan: the war with Armenia, the massive de-industrialization of the 1990s, and the oil boom of 2003, which gave rise to a skyrocketing growth of the Azerbaijani economy but also heightened growing social inequalities and poorly calibrated rent redistribution.

Thanks to this presence in Murmansk since the 1980s, today Azerbaijanis form a kind of "aristocracy" among migrants. Several success stories confirmed their *primus inter pares* status as a model of integration to follow. Gradually in the 1990s and the early 2000s, the small kiosks (*lavki*) selling fruit and vegetables were transformed into supermarket chains. Today, some major Azerbaijani families head up large networks of supermarkets, such as Evroros, and have succeeded in opening some 30 stores in other towns of the peninsula.[23] Most of these Azerbaijanis obtained Russian citizenship in the 1990s; they are no longer aliens legally, socially, or culturally. This first generation speaks good Russian, their spouses have joined local professional circles, their children are far more attached to Russia than to Azerbaijan, and they only see Azerbaijan as the place they will go for retirement. Many became ethnic entrepreneurs, helping new arrivals from their home region and playing a crucial role as mediators or brokers between migrants and Murmansk's authorities and employers.

For the new generations of Azerbaijanis who arrived in the post-Soviet period, social status is more tenuous and identities are shifting: some have Russian citizenship but would prefer to return and work in Azerbaijan, others have held onto their Azerbaijani passport and have a work permit in Russia, still others are migrants with a simple temporary registration, and some are illegal. From a socio-professional viewpoint, many of them work outside their area of qualification, despite receiving higher education in Azerbaijan. Others, particularly at the markets, are poorly educated and therefore more exposed to the fluctuations of the labor market.[24] Nevertheless, coming to work in the Russian Far North also makes sense as a way to learn new skills on the job. Given the tremendous level of youth unemployment in Azerbaijan, the Russian Far North's image as a worker's paradise remains highly enticing.

Murmansk's more vulnerable migrants: the Central Asians

The sequencing of generations and categorization of ethnic niches is quite different for Central Asian migrants. Central Asian immigrants in the Russian Far North cannot rely on a first generation who arrived during Soviet times for their military service: many Central Asians did their service in their home republic. Unlike the Azerbaijanis, they cannot rely on an existing socioeconomic and institutional base, and they are more vulnerable to the labor market given their lack of a history of mobility in the Far North.

Central Asian migrants seem more temporary than those from Azerbaijan. Large numbers of Central Asians want to go back to their native republics and are not interested in getting Russian citizenship.[25] The majority of them have spouses and children back in their home village, eschew Murmansk's social scene, and focus on their work to maximize the amount of the remittances they send to their families. Their degree of integration in the Murmansk social fabric is therefore weaker. Their fluency in Russian is often poor, and those who work in teams led by a member of their own ethnic group may live in an almost entirely Uzbek or Tajik environment without speaking any Russian. Lastly, their qualifications are often minimal, which limits their job prospects to the low end of the social ladder.[26]

Central Asian migrants are concentrated in several well-defined economic niches. Tajiks are established as a workforce at PEK, a lucrative transport logistics company that moves furniture all over Russia, as well as to China and to Kazakhstan.[27] They are regarded as serious, prudent workers who do not drink.[28] The Uzbeks have moved into the fruit and vegetable kiosks vacated by Azerbaijanis; this niche has been transferred from an increasingly well-integrated Azerbaijani diaspora to new waves of migrants. The Uzbeks are also employed in Azerbaijani supermarkets – the men as warehousemen and women as cashiers. Some Uzbek migrants also work for Nord West, a Russian fish-processing company that allegedly recruits via employment agencies in Bukhara and Navoy and sends Uzbek workers over in teams.[29] Lastly, the PhosAgro complex in Apatity and Kirovsk has begun to develop shift rotations (*vakhtovyi metod*), bringing teams of Central

Asians to work for two to three months at a time.[30] The small settlement of Koachva, close to Apatity, which was totally depopulated in the 1990s, today hosts barracks for these long-distance commuters from Central Asia.[31]

Migrant social networks and integration processes in Murmansk

The strategies implemented by Azerbaijani migrants to gain access to the labor market in Murmansk rely on the existence of a longstanding and institutionally established diaspora under the guidance of Afil Guseinov, the representative of the branch of the Azerbaijani Congress of Russia (*Vserossiiskii azerbaidzhanskii kongress*, VAK), created in Moscow in 1998 and in 2004 for the Murmansk region. Born in Nakhichevan in 1966,[32] Guseinov is part of the first generation of newcomers who did their military service in the Kola Peninsula and then decided to stay. The association plays a critical role in giving administrative aid to Azerbaijani migrants – including Russian language courses and legal support – and in facilitating economic and social integration, such as putting them in contact with employers, providing security deposits so they can rent apartments, and so on.[33]

The Central Asian diaspora association, Aziya, was established in 2008 by Paridokhon Nasirova. Born in Andijan, Uzbekistan, in 1960, Nasirova is one of the rare Central Asian newcomers of the first generation, a trajectory that can be explained by her marriage to an Azerbaijani who did his military service in the Northern Fleet. Aziya helps migrants from Central Asia to adhere to Russian migration policy, provides courses in Russian to prepare them for exams, gives advice on difficult family or medical situations, and visits prisoners and migrants in temporary custody. Thanks to a grant from the Russian Labor Ministry, the association is supported by two social workers and a lawyer.[34]

The extensive network of diaspora associations offering services and support to the newcomers provides a social safety net and makes it possible for migrants to manage administrative, financial, and personal difficulties. These associations make use of informal social networks such as family ties and generational and intergenerational solidarity to gain access to the local labor market and to facilitate migrant integration. These elaborate mechanisms work to prevent "social insecurity" (Castel 2003, 95) and confirm the high level of symbiosis between economic exchange and social relations (Granovetter 1973, 1985). Studies done among migrants applying for registration in Murmansk city show that they arrived with established connections: either they have a family member located there, who will help with finding housing and employment, or they belong to a group (often a village or an extended neighborhood) that has connections with a local employer. The latter also rely on these informal networks and hire only migrants who are endorsed by those already working in the firm.[35]

In the near future, new forms of socializing will likely emerge for migrant communities. The diaspora associations have not – for several reasons that are outside the scope of this chapter – erected a mosque in downtown Murmansk and had to compromise on a smaller religious house in the city's suburbs, but religious institutions are destined to play a greater role in identification and integration processes, as is already the case in many other Far North cities in Russia.[36]

Migrant identities offer a complex interplay among several layers of refer-ents. If migrants from Azerbaijan and Central Asia keep dense links with their home country, and often project themselves back "at home" for their older years, the emotional links to Murmansk are powerful, too. Interviews organized with migrants taking Russian-language proficiency tests reveal that at least one-third of them consider themselves to be long-term migrants in Russia and aspire to integration. They select the most complicated test that gives them a right-to-work permit good for three to five years, not the simpler one that gives access to a *patent* (license) for few months. Migrants often celebrate Murmansk as a city with little ethnic conflict, especially compared with some other Russian cities (Razumova 2004, 9). Migrants regularly cite the fact that employers often help them to integrate and the relatively low level of corruption of law-enforcement agencies in charge of migration as reasons they chose Murmansk as their destination.[37]

Murmansk has retained its Soviet-era reputation as the "frontier," and many migrants internalize this identity. For them, having a job in Murmansk is a sign of professional recognition – even if they work in sectors that do not correspond to their education – and of success, not only financial but in terms of social capital. Migrants internalize the local narrative about "Northerners" (*severyane*) being honest, work-oriented people, able to survive in harsh climate conditions thanks to strong ethics and solidarity. Symbolic integration, as critical as material inte-gration, seems therefore in process, helping to anchor migrants into the Russian Arctic's grand narrative of the pioneer frontier.

Conclusion

This brief sketch of Murmansk city and its population movements reveals sev-eral elements. In many aspects, Murmansk is a typical city for Russia's Far North, facing widespread socioeconomic transformations and demographic decline after a "Golden Age" under the Soviet Union. Yet, Murmansk also offers some unique features that differentiate it from its counterparts. Its location, close to Norway and Finland, but also its good connections with Russia's main European cities, as well as a diversified economy, give it advantages in attract-ing new populations. The Far North has always been, and will remain, a region of high mobility, with residents staying there mostly during their working life and leaving for more favorable climates at retirement. It has entered a new phase of development, in which population movements are no longer shaped by the state, but by a plurality of private and public actors who have to compete to attract labor. Murmansk combines state-sponsored economic sectors, such as the Northern Fleet, with private sectors such as the fishing industry, Arctic trade, and cross-border activities. While the Murmansk population has a high turnover rate, it still offers a certain urban identity that rubs off on newcom-ers from Azerbaijan and Central Asia. New immigrants absorb Murmansk's "Northerner" identity and find ways to integrate into the city's social fabric. These processes, so far understudied by Arctic specialists, highlight the Arctic as a globalized region, in which social and cultural changes occur at a rapid pace, and which call for more comparative studies.

Notes

1 "Nordic benefits" were comprised of three different systems: the "local coefficient," calculated by region – Murmansk was in the highest category, which almost doubled the salary, the "Nordic allowances" (*poliarki*), calculated by region but also by the number of years someone already spent in Arctic or sub-Arctic regions, and the "Nordic holidays," which added several more weeks of vacation paid by the employer.
2 See "Demograficheskoe razvitie i etnicheskie protsessy," *Kol'skaia entsyklopediia*, no date, http://ke.culture.gov-murman.ru/murmanskaya_oblast/5238/.
3 Ibid.,102.
4 Interview with local researchers working on the city's economic development.
5 Student population collapsed by one quarter between 2009 and 2012. Ibid., 36.
6 "Too expensive: Gazprom puts Shtokman on hold," *Russia Today*, August 29, 2012, www.rt.com/business/shtokman-gas-gazprom-857/.
7 See "Administratsiia Severnogo morskogo puti," at http://asmp.morflot.ru/ru/razresheniya/. I thank Sergey Goncharov for providing me with this data.
8 The eight monocities of the Murmansk regions are Kirovsk, Kovdor, Zapoliarni Zori, Monchegorsk, Tumannyi, Zapoliarnyi, Revda, and Nikel.
9 "Kadry dlia Severa. Noril'skii Nikel' preodolevaet defitsit kvalifitsirovannogo personala," *Izvestiia*, August 29, 2015, http://izvestia.ru/news/519193.
10 This analysis is the result of several interviews done in Monchegorsk, Nikel, and Zapoliarnyi with local researchers or administration representatives.
11 There are currently two collective Saami farms, with less than 30 people involved in seasonal transhumance, and a declining reindeer population, which in 2013 was as little as 40,000. Interview with Ivan Golovin, one of the directors of the Saami village, July 10, 2015.
12 2010 National Census, Rosstat, www.gks.ru/free_doc/new_site/perepis2010/croc/perepis_itogi1612.htm.
13 "Strategiia sotsial'no-eokonomicheskogo razvitiia Severo-Zapadnogo federal'nogo okruga na period do 2020 g," November 18, 2011.
14 "Upravleniia gosudarstvennoi sluzhby zaniatosti naseleniia Murmanskoi oblasti o dvizhenii rabochei sily na rynke truda Murmanskoi oblasti v ianvare-mae 2015," *Komitet po trudu i zaniatosti naseleniia Murmanskoi oblasti*, June 2015, www.murman-zan.ru/Attachment.axd?id=2b3df3aa-c84e-4926-9045-1536ff7dd81f.
15 *Federal'naia sluzba migratsii Murmanskoi oblasti*, information collected by Alexandra Bursteva.
16 2010 National Census, Rosstat, www.gks.ru/free_doc/new_site/perepis2010/croc/perepis_itogi1612.htm.
17 These numbers are the ones given by diaspora associations, and include both those who get Russian citizenship, long-term and short-term registration, and illegal migrants.
18 Interview with Afil Guseinov, Murmansk, July 8, 2015.
19 "Itogi vserossiiskoi perepisi naseleniia 2010 g.," *Rosstat*, http://murmanskstat.gks.ru/wps/wcm/connect/rosstat_ts/murmanskstat/ru/census_and_researching/census/national_census_2010/score_2010/ 1
20 Interviews with Azerbaijani workers in Murmansk markets, July 2015.
21 Interviews with Azerbaijani diaspora members, Murmansk, July 2015.
22 Interviews with Azerbaijani diaspora members, Murmansk, July 2015.
23 Interview with Afil Guseinov, Murmansk, July 8, 2015.
24 Interviews with Azerbaijani diaspora members, Murmansk, July 2015.
25 Interview with Paridokhon Nasirova, Murmansk, July 13, 2015.

26 Interviews with Uzbek and Tajik diaspora members, Murmansk, July 2015.
27 See their website pecom.ru.
28 Interview done by Alexandra Burtseva with PEK administration, Murmansk, June 17, 2015.
29 Interview with Paridokhon Nasirova, Murmansk, July 13, 2015.
30 Interviews with (anonymous) representative of the Montchegorsk Combine, July 10, 2015, and Zapoliarnii-Nikel on July 14, 2015.
31 Interview with an (anonymous) geologist, Apatity Institute of Geology, July 9, 2015, and with scholars at the Apatity Center for Social Questions of the Barents region, July 9, 2015.
32 The Nakhichevan region is renowned for its elites, and traditionally provided intellectual and political elites both for the Azerbaijani states and for the Azerbaijani diasporas.
33 Interview with Afil Guseinov, Murmansk, July 8, 2015.
34 Interview with Paridokhon Nasirova, Murmansk, July 13, 2015.
35 These findings emerged from the many interviews done by Alexandra Burtseva with Murmansk's main migrant employers, Murmansk, 2013–2015.
36 Interview with Bakthior, imam of the Murmansk prayer house, Murmansk, July 13, 2015.
37 These findings emerged from the interviews done by Alexandra Bursteva with migrants coming to take the test for registration, Murmansk, 2013–2015.

9 Ugra, the Dagestani North

Anthropology of mobility between the North Caucasus and Western Siberia

Denis Sokolov

The Khanty-Mansiysk and Yamalo-Nenets autonomous districts (hereinafter, respectively, KhMAO and YaNAO) are the main oil and gas producing regions of Russia. Unlike many other regions in Siberia, they have been able to maintain positive social and economic dynamics that favor their demographic growth, and thus they do not face the same depopulation patterns that plague many cities to their east and north. In both regions, indigenous peoples comprise a very minor part of the local population: Khanty and Nenets number about 30,000 people each, and Mansi less than 12,000. The majority of the population is Russian or belongs to the Russian-speaking Soviet world – Ukrainians and Byelorussians. However, a growing segment of the population also comes from more southern regions, especially from Azerbaijan – Baku has engaged in oil extraction since the early twentieth century, and many Azerbaijani engineers have found work in the big Russian oil and gas enterprises since Soviet times – and from the North Caucasus, especially Chechnya and Dagestan.

In this chapter, I focus on the Dagestani migrants living in KhMAO and YaNAO. Dagestan has an unusually high level of linguistic diversity – several dozen ethnicities coexist, although there are important cultural gaps between high-landers and plains dwellers – and a sharply declining socioeconomic situation. Unemployment has increased dramatically since the liquidation of the *kolkhozs* (collective farms) in 1992 (Karpov and Kapustina 2011), and lowlanders, who were accustomed to a relatively high standard of living, faced a dramatic degradation of their economic situation with the rapid collapse of the local rural economy. This created several "push" factors for the mass emigration of some ethnic groups, mostly Nogais, Kumyks, and Azerbaijani Lezgins, to other parts of Russia.

Dagestani migrants have clustered in certain sectors of the Russian economy, particularly construction. They predominated in megaprojects such as the 2014 Olympic Games in Sochi, the APEC (Asia–Pacific Economic Cooperation) summit facilities in Vladivostok, and the multiple massive residential and commercial real estate projects in Moscow.[1] Other professional niches taken by these migrants include markets and bazaars – not just the famous Cherkizovsky, Luzhniki, and Lyublino[2] in Moscow, but also in all Russian main cities – as well as freight and passenger transport, taxis and car service stations, and truck driving (*dal'noboishchiki*). Since the late 1960s, tens of thousands of pastoralists from Dagestan have gradually relocated to the eastern Stavropol region, and today thousands of seasonal migrants (*gektarshiki*), mainly Tsumadins and Kumyks, rent one or more hectares of irrigated land for growing and selling onions, garlic, herbs, vegetables, and melons in the Northern Rostov region, Astrakhan, Volgograd, and Kalmykia. Finally, many Dagestani have penetrated the private security sector and the armed forces. Hundreds, if not thousands, of young people from Dagestan returned home as "Cargo 200" (dead corpses) riddled with gunshots and stab wounds, from Moscow, St. Petersburg, Rostov-on-Don, and almost all other regions of the country.

Many rural youth from Nogai and Kumyk villages have opted for a mass exodus into pre-fabricated towns and the dormitory-type barracks of the Tyumen region in search of work. These newcomers from Dagestan, even with their Russian passports, rarely are treated as full-fledged citizens. Difficulties with registration, housing, and job placement, on top of harassment by law-enforcement agencies, leave many migrants outside of the formal public space. Only the second generation of migrants, raised in KhMAO and YaNAO and with better local connections, can gradually enter previously forbidden sectors such as healthcare, education, public administration, and the oil and gas industry.

Migration from the North Caucasus to other Russian cities, especially to Western Siberian oil and gas cities, has generated few academic studies. The topic is politically sensitive, and it is difficult for external observers to access first-hand data and win the trust of migrants, who are accustomed to being wary of strangers asking questions and prefer to rely on regional and kinship solidarity. After several years of field work in Dagestan, I was able to overcome many of these barriers. This study was conducted in several cities of the Khanty-Mansi district: in Surgut, the regional capital and home base of the oil company Surgutneftegaz; in Novaya Fedorovka, an industrial town near Surgut, whose Fedorovsk oil field is being developed by Surgutneftegaz; in the small settlement of Ult-Yagun, near the railway station adjacent to the Fedorovsk deposit; in Russkinskie village, built for the Khanty, the local indigenous people; in Nefteyugansk, the former headquarters of Yukos oil company, which now belongs to Rosneft; in Nizhnevartovsk, the headquarters of the oil company TNK–BP; in the town of Pyt-Yakh, founded to develop the Mamontovskoe field; and, finally, in Tyumen itself.

This chapter presents a kind of anthropology of mobility in these sub-Arctic regions of Russia. After examining the migration patterns in Tyumen region, I explore the ethnic clustering of many of these oil and gas cities and the parallel lives built by migrants there. I then study the mobility mechanisms put in place by migrants, who develop parallel lives in two locations: what they call the "mainland" (*zemlia*, "earth"), i.e., their homeland, and the "North" (*sever*), the KhMAO and YaNAO districts. The term *Ugra* (the Russian abbreviation for KhMAO), now commonly used in Dagestan, has become a synonym for this new, second life built by migrants far north of their home towns, in a totally different natural and cultural environment. New forms of connectedness between the mainland and the

North have structured in terms of work, education, healthcare, and leisure, as well as in terms of family development and religious life.

Migration patterns in the Tyumen region

Migrants are attracted to Russian regions with a large GRP (gross regional product), typically Moscow and Moscow oblast, St. Petersburg and Leningrad oblast, Tyumen, and Krasnodar. Tyumen oblast accounts for 91 percent of the total Russian production of natural gas, 66 percent of oil, and 9 percent of electricity, and its GRP per capita is the highest in the Russian Federation. Thanks to this economic specificity and the revenue it generates, the Tyumen region is among the leading immigration regions, competing with St. Petersburg, Moscow, Moscow oblast, and the Krasnodar region.

Tyumen experienced massive urbanization in the 1960s, with the discovery and development of oil and gas fields there. Since 1965, the region's population has doubled every ten years. This trend stabilized only in the 1990s, when job-seeking Russians reoriented toward the country's more European regions. The oblast is therefore an extraordinary crossroads. Migratory outflow is significant, especially among the elderly population who head for milder climates when they reach retirement age, but also among youth, who leave to pursue their education, and even for working-age people. These transit flows are largely defined by ethnicity: essentially, ethnic Russians and Slavic populations leave, while migrants from the North Caucasus and Central Asia arrive (see Table 9.1).

Table 9.1 Dynamics of changes in the ethnic composition of Tyumen oblast

Population of Tyumen oblast			
	1989	*2002*	*2010*
Total	3,097,657	3,264,841	3,395,755
Did not mention	219	31,240	187,803
Avars	0	2,766	3,783
Dargins	0	2,663	3,722
Ingush	0	2,391	2,994
Kumyks	3,854	12,343	18,668
Lezgins	3,988	10,631	16,247
Nogai	1,039	4,272	8,888
Chechens	4,587	10,623	10,502

(continued)

Population of Tyumen oblast

	1989	2002	2010
Bashkirs	41,059	46,575	46,405
Belarusians	49,057	35,996	25,648
Kazakhs	15,682	18,639	19,146
Russians	2,248,254	2,336,520	2,352,063
Ukrainians	260,203	211,372	157,296
Azerbaijanis	19,455	42,359	43,610
Others	817,300	821,195	763,039

Source: Rosstat, census data.

The main influx of migrants from the North Caucasus took place in the 1990s and early 2000s. The reduction in revenue in the oil and gas industry, due to falling oil prices and the redistribution of assets, as well as dwindling air travel opportunities, led to a major transition among oil workers. Many stopped working by rotating shifts and switched to permanent residence in the KhMAO and YaNAO. Validation of this can be found in census data and in my field materials, collected in Surgut, Pyt-Yakh, Nizhnevartovsk, and Nefteyugansk between 2011 and 2014. Life in KhMAO and in YaNAO is built around oil and gas deposits and is organized by extraction companies. Migrant settlements are thus, to a large extent, dependent on job distribution patterns.

Natives of the North Caucasus republics have settled in all regions of KhMAO and YaNAO, but in some places they live in compact settlements. In the Surgut region, the small town of Novaya Fedorovka, built for Komsomolskneft and Surgutneftegaz subsidiaries, features about 9,000 North Caucasians (approxima-tively 5,000 Nogai, about 1,000 Dargins, a similar number of Lezgins and Avars, and other ethnic groups in smaller numbers), so an important part of the 22,000 total population. Buses leaving Kizlyar, in Dagestan, for KhMAO are called *fedor-ovskie*, as the majority of migrants are going to Novaya Fedorovka. The Nogai domination is such that newcomers from Uzbekistan try to pass as Nogai in order to get jobs – very often car services – and housing (author's field notes, Novaya Fedorovka, 2011, 2014). A relatively similar compact settlement can be observed in Vertoletka-aul in Pyt-Yakh, opened in 1965 as a dormitory-type barracks village for exploiting the Mamontovskoye oil field, and where several thousand Kumyks live today. Divnoe village in Nizhnevartovsk is a third example of a self-contained Kumyk settlement, with more than 3,000 of them.

However, the exact number of Dagestani is difficult to calculate, because many of them do not register with state services and live without a valid residency registration (*propiska*). Many circulate regularly between Dagestan and the North, and therefore are officially registered in their home region. To try to get a more precise statistical picture, I used data collected by the leaders of the national cultural autonomies, a type of nongovernmental organization registered by the Russian Ministry of Justice since the end of the 1990s that represents ethnic minorities living in compact settlements. National cultural autonomies are able to collect demographics based on lists of *auls* (the traditional village). They raise money to send home the bodies of the dead countrymen, distribute wedding invitations, and organize events, such as football competitions (author's field notes, Surgut, Novaya Fedorovka, 2014).

The gap between official and unofficial data is large (see Figure 9.1 and Table 9.2). For instance, Gubkinsky (YaNAO), with over 26,000 inhabitants, officially has only 151 Dargins, while expert and peer evaluations calculate more than 900 (author's field notes, Surgut, 2014). In the whole Tyumen region (including KhMAO and YaNAO), if one adds up the expert estimates, there are about 50,000 Lezgins, almost 40,000 Kumyks, and between 25,000 and 30,000 Nogais.

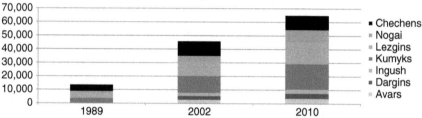

Figure 9.1 Ethnic composition of certain categories of population of Tyumen oblast.

Source: Rosstat, census data.

Table 9.2 Comparison of official (2010 census) and expert (field research carried out in 2011–2014) estimates of migrants from the Caucasus to Tyumen oblast

Population of Tyumen oblast

	Census of 2010	*Expert evaluation of 2014*
Avars	3,783	7,566
Dargins	3,722	7,444
Ingush	2,994	5,988
Kumyks	18,668	40,000
Lezgins	16,247	50,000

(*continued*)

Population of Tyumen oblast

	Census of 2010	*Expert evaluation of 2014*
Nogais	8,888	30,000
Chechens	10,502	21,004
Russians	2,352,063	2,352,063
Azerbaijanis	43,610	300,000
Others	763,039	763,039
Central Asian region		100,000

According to Ahmed Yarlykapov, an anthropologist from the Miklukho-Maklai Institute of Anthropology and Ethnology in Moscow, "If in the 1990s migration was relevant only to those who had no job and livelihood at home, in the 2000s, there is a noticeable group of migrants who move to the North in search of earnings higher than at home. Even those who, by local standards, get quite a decent salary (5,000–6,000 rubles a month) and have a stable job, predominantly state employees, tend to migrate" (Yarlykapov 2008). Many of our informants noted that the peak of migration took place in the years of 2000–2003, which coincides with the beginning of a rise in oil prices and higher standards of living for Western Siberian inhabitants (author's field notes, Surgut, Novaya Fedorovka, Nizhnevartovsk, 2014).

The main migration wave of the second half of the 1990s to the mid-2000s has partly exhausted the Dagestani plains' workforce. Almost everyone who could and wanted to move has already gone. Migrating, permanently or on a rotational basis, has become mainstream – a routine matter.

> When I was leaving in 1996, there [on the Russian mainland] … the whole village would see me off, as if I were leaving for the army. We had a big feast, and everyone came to say goodbye "you are leaving for the North." My sister said, "That's it, you are leaving … you will get Russianized, you will forget everything … everyone will get lost." Then we left, and after me, my brothers arrived, two of them, then my sister came. … She lives in Urengoy. And now there are more Nogais here than where we used to live. (author's field notes, Surgut, 2014, Moscow, m., 50 years old)

Moreover, the emerging occupational niches are now being filled by a second generation of migrants: Now "70 to 80 percent of young people are those who were born here or arrived here as children. They are now 30 years old, and their children are ten" (author's field notes, Surgut, 2011 T., m., 55 years old).

Unlike workers from Central Asia, who tend to come back home infrequently because it might endanger their legal situation if they are working illegally, Dagestanis are citizens of the Russian Federation and therefore free to circulate throughout the country. Many of them, after years of work in the Tyumen region, return to their home region, or to Stavropol or the Caucasian Mineral Waters region. Some also go to Tyumen itself, which is considered a better place for retirement or education purposes than the oil cities, and more attractive than the place of origin.

> Nowadays, some people got plots near Tyumen, almost in the city, they are building houses, private houses. There are some from our village. One guy went, bought the plot, built the house, and doesn't want to go back to Dagestan; he wants to stay in Tyumen. He says, the children are here, there is land, there is a normal climate in Tyumen. … Here [in Surgut] the land is just so expensive, not everyone can buy it. For example, six-hundredths of a hectare costs 800,000 rubles. And then this is swamp land, you have to drain it. (author's field notes, 2014, m., 50 years old, Nogai, from Rassvet village in Kizlyar district of Dagestan, has lived in KhMAO since 1997)

Globally, those who can afford also leave the smaller cities of KhMAO, for instance Pit-Yakh, and go to the region's bigger cities, where services are more developed. "Pit-Yah is left not only by Kumyks, but by all the young guys, no matter what nationality, if they find a more or less normal job somewhere, they leave" (author's field notes, MG, m. 46 years old, Pyt-Yakh).

Parallel worlds: ethnic clustering in Northern cities

Because of these massive population flows, Russia's oil and gas cities have developed as complex, multi-ethnic, and clustered societies. Migrants from the North Caucasus themselves have often maintained their family and kinship structure, *teip*-based or *jamaat*-based, which sometimes has to assume responsibility for security, protection of property, as well as other economic and political interests of its members. Ethnic discrimination is especially noticeable in employment, the provision of public services, and strained contacts between migrants from the North Caucasus and Central Asia and local law-enforcement bodies.

The relationship between North Caucasians and law-enforcement bodies in several regions of the Russian Federation is often reduced to illegal detention and arrests during mass brawls. The police are not perceived as a protective institution; on the contrary, they are associated with arbitrariness and corruption. The Dagestani culture of conflict resolution eschews law-enforcement practices. Natives from the North Caucasus are less willing to turn to the courts for justice and prefer to use personal relationships with employees of the law-enforcement bodies, local kinship associations, and sometimes connections in the criminal world to solve problems. In everyday life, law-enforcement agencies tend to harass North Caucasians as if they are foreign migrants, even though they are Russian citizens. For instance, in KhMAO and YaMAO, it takes over six months to re-register vehicles originally registered in Chechnya, a timeframe that is drastically longer than transfers from other regions of Russia.

Interethnic tensions frequently arise. In 2014 in Surgut, a crackdown on modified Prioras (a new version of the Lada car produced by AvtoVAZ)[3] resulted in mass protests, and reports of police officers beating Caucasians (author's field notes, Surgut, 2014). Officials allegedly extorted bribes from entrepreneurs, one of whom was killed near a tire shop.[4] Far-right organizations and skinhead groups are well developed in the region, sometimes with the tacit support of police officers. The response of North Caucasians to this state brutality and xenophobia is mass mobilization, radicalization of youth, car stereos blasting *lezginka*, the national music of Dagestan,[5] as well as stickers, signs, and T-shirts emblazoned with "Dagestan" (Kapustina 2014).

To reduce interethnic tensions, the Russian authorities created Coordination Councils for National Policy at the municipal and district levels. These bodies are supposed to organize discussions, lectures, and workshops on "tolerance and integration" and coordinate with the heads of national cultural autonomies. Thanks to this cooperation channel, the Surgut municipal authorities agreed to stop requiring a photo and fingerprints when registering Russian citizens from the North Caucasus at the passport office (author's field notes, Surgut, Novaya Fedorovka, 2014). However, this bureaucratic hurdle has not been successful in many aspects. Informal mechanisms function better, hence the widespread insistence by the local population that North Caucasians contribute to corruption because of their willingness to pay for access to public services.

Dagestani in the Tyumen region have long been excluded from the local real estate market. Stigmatized as "persons of Caucasian nationality," (*litsa kavkazkoi natsional'nosti*) it is challenging for them to rent an apartment, and they have to accept an overpriced room in a dorm, where they live with their entire families. In some cases they "buy" an employment contract from the departmental housing officials, including fellow countrymen, at a price comparable to the cost of a full apartment. However, since the mid-2000s, the housing situation has changed for the better. District housing subsidies for children, in the amount of 50 rubles per square meter, until recently allowed receiving up to 600,000 rubles/year per child, which could be added to the motherhood bonuses given by the state as part of a pro-natality program. Now more established in the region, some Nogais have even opened real estate firms, joking they are working only for their community: "Slavic people, do not bother calling" (author's field notes, Surgut, 2014, Moscow, m., 30 years old). Many Dagestani who were able to get two or three apartments are now renting them to their ethnic kin or to migrants from Central Asia, creating a kind of parallel housing market for newcomers.

Discrimination in employment remains a major challenge for people from the North Caucasus. It is almost impossible for them to get a job with the police, prosecutor's office, Ministry of Emergency Situations, or anything related to justice and tax collection, even with formal qualifications. Even some more minor positions are difficult to get, such as seller at a Surgut-based household chemicals factory:

> She was successfully interviewed and came back with a passport, gave them the passport, and they saw Grozny, "No, no, – we cannot hire you" – they said. "Why didn't you tell me in the beginning?" – she asked. They said: "But we did not know that you were a Chechen. What do you want from us?" (author's field notes, Surgut, 2011, T., m., 60)

Employment is therefore almost universally arranged with the help of fellow countrymen. Getting a job in the oil and gas industry, such as driller or assistant driller, is considered both prestigious and financially rewarding (around 150,000 rubles per year). At Surgutneftegaz, for instance, contracts often guarantee employment to children whose parents are employed by the company. With this in mind, some Nogais have been able to get hired as drivers or paid bribes to become assistant drillers, giving their children, after receiving an education in the oil sector in Tyumen, the opportunity to be hired as drilling masters. In Novaya Fedorovka, the "capital" of Nogais in KhMAO, in the past there was a "special fee" of 300,000 rubles that a Dagestani had to pay to be employed in the oil and gas industry, an expensive but profitable long-term investment to secure their children's future.

The majority of Dagestani thus have to target other professional sectors. Those with a higher education degree and good work experience, such as doctors and surgeons, leave Dagestan in search of better positions with higher salary,[6] or because they would like to work in a less corrupt environment. They can get a job in KhMAO cities if family members will help them secure a local residency registration, a mandatory document to get hired. In Surgut, each of the nine public medical clinics employs several Nogai specialists.

To succeed economically, many Dagestani migrants work multiple jobs spanning the private, retail, and trade sector. In addition to their main job, such as working on a drilling station, they open a small auto repair shop – often considered to be an "Azeri specialty" – or work as driver for cargo transportation, livestock, and service companies, which is considered a "Nogai profession." Another way to cope with the job system is to open small shops that rent or lease specialized equipment to local businesses, especially the construction sector. Many migrants also decide to open a business in two regions simultaneously, for instance in the North and in Moscow region. Dagestani businessmen who succeeded economically unfailing give priority to their kin. For instance, Mahammadrasul Yusupov, a Dargin from Kaytag and one of the owners of Surgut Avtodorstoy, employs mostly Dargins, Lezgins, and Kumyks. Respondent G., a 46-year-old Avar from Orochi village in the Khunzakh district of Dagestan, owned a security company in Stavropol krai in the 1990s. He is now a contractor at complex engineering facilities for oil and gas companies and employs up to 60 fellow villagers at a time. On a regular basis there are 10–15 people from his native village working in his firm.

These clustered businesses cannot be explained only by an "ethnic" factor, and in fact combine multiple types of solidarity, such as family ties, business colleagues, and interpersonal trust. These "networked" businesses allow for the transfer of skills and knowledge between compatriots, help to reduce the costs of going into business, guarantee the safety and protection of property, and enhance collective reputations (Tirole 1996).

The structuring of associative life promotes integration into the new environment but also reinforces the "parallelization" of lives and the social dynamics between Dagestan and the oil and gas cities of KhMAO and YaNAO. For some groups, such as Chechens, maintaining social ties among the migrants was a priority long before the birth of an official association, and group consolidation has been tightly linked to commemorative practices:

> We have a day of mourning, February 23,[7] we observe it. And we celebrate it as the day of our diaspora. But, in fact, we were established as a fraternity, formed after we held a day of mourning for the first time. Although we talked before that too, this was an official gathering on this occasion. … It was sometime in 1989. But back then we had few members of the organization. (author's field notes, Surgut, 2011, T., m., 65 years old)

Almost everywhere, Chechen national groups are combined with the Ingush, under the encompassing ethnonym of Vainakh. For the Dagestani, national cultural autonomies are less important and reduced to the organization of cultural and sporting events such as football championships between the *auls*, organized by Nogais in Novaya Fedorovka, participation in town festivals, organization of national ensembles, like "Lezginka" in Tyumen, and membership in the Coordinating Councils of the local administration. One of the key roles of the national autonomy associations is to take care of fellow countrymen, especially sending dead bodies back home.

> If there is a diaspora organization that represents Dagestan, it must first of all have a phone in the morgue. If something happened to a countryman he [the representative of the organization] needs to be reached by phone and arrive there. And this organization should have the funds that it collects, in order to do everything and send the person safely home. Cultural and leisure [activities] are of secondary priority. (author's field notes, Surgut, 2014, G., m., 46 years old)

Dagestani national groups rarely collect money to support fellow citizens in need; this is mostly done through the mosque. *Zakat* (the alms-giving mandatory in Islam) is regularly collected by the imam. In Pyt-Yakh the local imam virtually combined the functions of a religious representative and a representative of residents of the Vertoletka-aul district, which is home to most of the city's Kumyks. The mosque established a funeral home, reads the *janaza* (funeral prayer), builds the coffins, and even bought a cargo vehicle for transporting bodies to the home-land. In Surgut, another system is in place: 30 Nogai families put funds together under the supervision of a respected 42-year-old athlete with four children, and sent money to build a mosque in their home village in Nariman.

Some national autonomies are under the leadership of an entrepreneur who takes on the role of patron and representative of his ethnic group in the Coordination Council of the city administration. The position of the head of the diaspora association, formal or informal, requires participation in all issues, and inevitably turns into a political role. For example, when fights between Chechens and Nogais lead to fatalities, people turn to the diaspora heads, who have to find solutions satisfactory to all the parties in conflict. In the case of a deadly conflict between Dargin and Azerbaijani truckers,[8] both heads of diaspora were asked "to resolve the issue" on behalf of the mayor of Makhachkala (author's field notes, Tyumen, 2014, V., m., 53 years old). The diaspora head therefore both cooperates and competes with religious figures, crime bosses, law-enforcement agencies, and the local Russian authorities. Another way of solving a conflict is "forwarding" it to the homeland, where customary law (*maslaat*, for example) or Shariatic leaders will be able to offer a resolution acceptable by all.

The Russian mainland and the North as a single habitable space

Many of the North Caucasian migrants are used to living double lives on two territories, the Russian mainland and the North. The dual pattern is well reflected in dozens of narratives that were recorded during my fieldwork in Dagestan, Ugra, and Tyumen. It is manifested through several social practices that I briefly describe here.

The existence of a single habitable space divided between two locations becomes very visible during the long Northern vacations that allow North Caucasians to come back to the mainland. In the summer, or sometimes for a longer period going from April to November, tens of thousands of vacationers from KhMAO, YaNAO, and Tyumen oblast take buses, cars, trains, and planes to rush back home to North Caucasus republics and to Azerbaijan (author's field notes, Surgut, 2011 and 2014) to visit relatives, celebrate weddings, and some-times to observe the Ramadan fasting if the calendar allows. The license plates of Surgut, Novyi Urengoy, and Tyumen vehicles are referred to as "golden" (*zolotoi*), as they are targeted for bribes by corrupt traffic police. Since a wedding is often one of the main occasions for these trips, and in some villages the majority of young people work in Tyumen oblast, sometimes all cars in the wedding proces-sions in Kaytag or Nogai districts of Dagestan have Surgut license plates. This is also due to the fact that many migrants bring vehicles from the North as gifts to the fathers, uncles, brothers, and nephews.

The wedding ceremony, a critical ritual of passage, has been adapted to this double location. Kumyks traditionally celebrate two weddings, one for the groom and one for the bride, over two days. The Northern work schedule amends the tradition: these two days are celebrated with an interval of one week in order to allow the family to travel from Dagestan to the North, while close relatives who work far away also go back to the home village for the first part of the wedding. Because of higher salaries in the North, the wedding gifts offered by each guest – and recorded in a special notebook – have been increased from 1,000–2,000 rubles (close relatives give up to 100,000) to 5,000 and even 10,000 rubles. Some marriages have a clear migration strategy at their basis, allowing a young man or a young woman to migrate to the North.

This life in two different locations is also visible in the housing strategies of families, who literally reside in two locations, as this 55-year-old resident of Novaya Fedorovka, an ethnic Nogai, a native of Oguzer village in Kizlyar district of Dagestan, explains:

> My brother has a house in Oguzer. I have a house in Novovladimirovka, I want to sell it, and buy one in Kizlyar, to move closer to my sister, she lives in Kizlyar. I bought a dacha[9] [in Novaya Fedorovka] and sold the apartment, I helped my son, and went to live in the countryside. My son is here too, also works in the organization... The son has an apartment in Fedorovka and in Surgut. ... We bought one entirely and for the daughter we took a mortgage, she is there with my son-in-law (also a Nogai), they have two children, and live and work in Surgut. (author's field notes, Surgut, Novaya Fedorovka, 2014, 2011)

Many migrants to the North hope to return home at retirement or even earlier. The first generation, those who first migrated at the age of 25, 35, and even 40, are those who dream of returning home, a strategy visible in the construction of new houses in their hometowns built through their remittances. The second and the third generations are less attracted by the notion of returning to Dagestan or Chechnya and prefer to reach Tyumen – or even Moscow and St. Petersburg. Retirement strategies have also evolved with the change of economic prospects: the reduction of earnings, loss of contacts in the home village, and children and relatives now based in Moscow or Tyumen encourage migrants to bring their family to the Russian mainland more than to return to their ethnic republic. As there are fewer people remaining back home, migrants have begun transporting elderly parents to live among their children and grandchildren in the North.

However, the dead are almost always sent home ("Sel'skie etiudi k gorodskomu peizazhu" 2013). With rare exception, such as the death of infants, Dagestanis do not bury their dead in the North. Instead, bodies have to be sent home, 4,000 kilometers away, in order to be buried following the religious or traditional customs. Throughout the years, sending corpses back home has become a model of mutual support; everything is organized quickly and in practical manner.

Mosques are also built in pairs: one at home and one in the North. In many Nogai settlements in Dagestan, new "Northern" mosques had been built, usually using migrant funds; a mosque in Novaya Fedorovka, for example, has been under construction for several years, with interruptions. Collected funds come not only from the Nogai diaspora, but also from other Muslim communities, including Dagestanis, Chechens, Ingush, Azerbaijanis, and migrants from Central Asia. Imams of Northern mosques point out that Dagestanis prefer to send money for a mosque in their native village, rather than donating to Muslim communities of Surgut and other cities around Ugra. Some Dagestanis explain this preference by stating that the home community should be helped before the migrant community:

> I say, dear Imam, what is written, who I should help first? … I have to help my parents, then my family, brothers, sisters, relatives, and then the village where I was born, and then, if I still have spare cash, then I can use it at my discretion. There are three or four, maybe five people like me in our village, the village is poor. You are here, Gazprom helps you, the city administration allocates money, Surgutneftegaz gives money, Tyumenenergo gives money, so there are some donations. (author's field notes, Surgut, 2011, G., m., 46 years old)

Being able to rely on this double space allows extended families to take what they consider the best of both systems and to play with the intersections of both cultures. For instance, migrants say that the level of corruption in YaNAO is less than in Dagestan. Therefore, they try to get medical treatment in Tyumen or Surgut and often register their mother or father in their Northern home, so they become automatically eligible for health insurance in the oblast. The same goes for those who need disability status, which cannot be obtained in Dagestan without paying bribes (author's field notes, Surgut, Pyt-Yakh, Nizhnevartovsk, 2014). Some Dagestani with a medical education build their careers in YaNAO and specialize in care for their North Caucasian fellows and Central Asian migrants. They sometimes work closely with other migrants selling "Islamic" healthcare products such as thyme oil, very highly regarded in Muslim medicine, and therefore build for themselves a specific, targeted, commercial niche. A similar situation occurs with education. Migrants invite their brothers and nephews to join them in the North, so they could go to colleges and universities there. They claim that a Dagestani diploma raises suspicions and xenophobia (author's field notes, Surgut, Novaya Fedorovka, 2011, 2014).

Multiple investments are made between the mainland and the North. Even if this kind of trade is often not recorded in official statistics, it shapes the everyday life of migrant communities as well as small and medium-sized private enterprises in both regions. Some migrants who succeeded in the North become involved in developing the economic prospects of their *jamaats* at home. G., a 46-year-old male Avar, has been living in Surgut since 1994; he created an agricultural enterprise in his native village, in the Orot Khunzakh region in Dagestan, in 2004. To generate income to pay the mortgage, he rented 150 hectares and even financed the restoration of rice crops on 50 hectares (author's field notes, Surgut, 2011, G., m., 46 years old). There are also widespread cases of reverse investments, from Dagestan to the North. Many business elites from the North

Caucasus, including retired criminal authorities, invest their earnings into a business in the North – mainly special equipment that can be leased to the high-tech foreign companies of the oil and gas sector (author's field notes, Surgut, Tyumen, Nizhnevartovsk, 2014).

Last but not least, success stories have made some migrants famous and respected in their home communities. This is the case of G., an Avar from Khunzakh district in Dagestan, a large contractor for technological construction in Ugra, who not only attempts to participate in the life of his *jamaat* directly or through his relatives, but also takes an active part in the local elections of his district. Z., a Nogai working in Novyi Urengoy, was an active participant in the protest against the construction of a sugar factory in his hometown in 2011 and supported direct elections for the district head (until then appointed by municipal deputies). As with several other Northern activists, he was encouraged to run for a local deputy seat. Many of these activists now dream of launching a Nogai Autonomy in Surgut and Novyi Urengoy that would allow them to develop parallel activities in several Russian towns. "I have a dream, to tell the truth, if we, God willing, establish a [Nogai] Autonomy, in any case we can create some institutes, universities, offices, in Stavropol or Moscow. We can establish [higher] education in the village, [in order] to work, to teach children" (author's field notes, Surgut, 2011, Z., m., 30 years old).

Regional trade has been reoriented by the massive flows of individuals and families regularly traveling between the Russian mainland and the North. First organized in private cars, the trade is now professionalized with several "gazelles" (light trucks) navigating between Terekli-Mekteb, Kizlyar, Novyi Urengoy, and Novaya Fedorovka. They transport meat, fish (Nogais popularized eating *kutum*, a fish, caught in Terek and Sulak Rivers, in Ugra), and fruits and vegetables, particularly watermelons. For a wedding they bring not only food, but also dishes and furniture from Khasavyurt and drive back home with appliances, used motorcycles, metals, and equipment such as valves and pipes.

The meat business is a very cost-effective one: in Dagestan meat is not sold for money; all transactions are carried out on credit, and are recorded in special notebooks. Many Nogais have specialized in sending large quantities of frozen meat to the North in order to sell it, at a higher price, and for "real" money, to Russians, but also to sell it in a more traditional, Dagestani way, to their kin. In 2006, one of my informants from Dagestan would visit his daughter, who was studying in Tyumen, once a month and bring 500 kilograms of fresh halal meat; he would sell it over the phone, out of his dacha: "Twenty minutes later, the meat was all gone" (author's field notes, Novaya Fedorovka, 2014, Moscow, m., 50 years old). In 2010, he stopped this business because of growing competition from other migrants who organized the transportation of larger containers of meat, about three loads every month and a half, in a five-ton refrigerated truck. In addition, there are now four farms in Beliy Yar, in Surgut district, where Dagestani raise and slaughter halal sheep from Altai and bulls from Omsk. On the outskirts of Surgut, the Druzhba market, which belongs to an Azerbaijani businessman chairing the local national cultural autonomy, is one of the main pockets of informal, migrant-based economy in the North, largely initiated by natives of North Caucasus. Many Lezgins, Dargins, and Chechens work there, and they have opened Caucasian cafes and halal shops that supply the whole Surgut region.

Finding Islam in the North

A generation gap seems to characterize migrants' perceptions of their Russian environment and their integration strategies. Children of first-generation migrants are schooled in the Tyumen region and therefore develop new sociability networks that are not systematically related to their ethnic or kin community. Some say children often hang out with their fellow countrymen at the request of their parents, who are concerned about the loss of Dagestani identity and traditions (author's field notes, Surgut, Novaya Fedorovka, Nizhnevartovsk). Some try to distance themselves openly from their ethnicity: an entire Lezghin *tuhum* (clan or extended family) based in Tyumen changed their last name to Tyumenovy, and a Nogai, working in Moscow as a graphic designer at a prestigious journal, took the pseudonym "Dima Fast" (author's field notes, Surgut, Tyumen, 2014). However, interethnic marriages are still limited: according to my informants, only a small percentage of young Dagestani marry Russian women who agree to convert to Islam, but most marry people from their ethnic group or from another Dagestani ethnicity.

Indeed, the dual identity built by Dagestani migrants in KhMAO and YaNAO is most evident in their relationship to Islam. It is often in the North that the migrants "come to Islam." Tyumen region is now considered to be one of Russia's centers of Islamic enlightenment facing the rapid spread of multiple religious denominations, including Salafi ones. Even if information is scarce, it seems that the Tyumen region has a high rate of ethnic Russians converting to Islam, based on attendance at the local mosques (author's field notes, Ult'-Yagun, Nizhnevartovsk, 2014). There are multiple ways of being Muslim. The majority of Dagestani define themselves as "simply Muslims" (*prosto musul'mane*) (author's field notes, Surgut, Tyumen, Novaya Fedorovka, Nefteyugansk, Pyt-Yakh, 2011, 2014) and do not want to identify with a specific denomination or trend, neither as Sufi nor traditionalist, nor Hizb-ut-Tahrir or Salafi.

The Tyumen region hosts more than 100 mosques, two-thirds of them situated in the southern part of the region, where Tatar and Bashkir villages are located. Tatars and Bashkirs were the first nominally – in the sense that individuals may not be believers much less practicing religion – Muslim population to be working on Western Siberia's oil and gas fields. Local mosques are thus generally under the administrative umbrella of the Ufa-based Spiritual Direction of Muslims led by Talgat Tadzhuddin. Migrants from Central Asia seem to integrate relatively well in these established mosques, and Tajiks, in particular, are recognized and celebrated as having a high level of theological knowledge compared with the local Muslims. Many of them serve as imams in the mosques (interview with the assistant imam at the Nefteyugansk mosque, 2014).

However, local observers and my own fieldwork confirm that the North Caucasian diasporas prefer to build their own mosques instead of joining the Tatar, Bashkir, and Central Asian ones. Nogais are followers of the Hanafi school, like many Central Asians, but still collect money to build their own community mosques both at home and in the North. Kumyks and other North Caucasians are mostly Shafii followers and they, too, have their own religious institutions which refer to the Spiritual Board of Dagestan, not to the Ufa one.

The migrant religious scene became more complex with the rise of so-called non-traditional groups, often with a proselytizing agenda such as the pietist movement Tabligh Jamaat or the more political Hizb ut-Tahrir. Local observers mention the rise of Salafi groups and preachers throughout KhMAO and YaNAO. The imam from the main Tyumen mosque, Galimzhan Bikmullin, in office since the 1980s, had to leave in 2014, reportedly for allowing several of his mosques to become places for Salafi propaganda.[10] Indeed, many young Dagestani and especially children from the second generation come to Islam not during their stay in their republic of origin, but in Ugra. They do not attend official mosques and prefer to join underground groups that meet in private apartments, under the leadership of a charismatic preacher. Several famous Salafi preachers have been traveling to the region to meet the local communities.

Those defending a more traditional Islam complain about the situation: "In Surgut several shops sell books by Albani and many others. Yes, you can find them in the book shops because nobody here checks or filters" (interview with an Islamic activist studying in a local medresse, Novaya Fedorovka, 2014, A., m., 30 years old). The local population also expresses some concerns about the lack of response from the state structures:

> KhMAO is not in the last place in rank of Russia's regions where extremist groups are growing. They develop very quickly here. Of course, we try to hamper them and to teach children not to visit terrorist websites. … But we don't have funding at the level they have. … Our services monitor this issue from far away. (author's field notes, Pyt'-Yakh, 2014, m., 48 years old)

These Islamic denominations represent more than a particular interpretation of the Quran and the Hadith; they are also linked to struggles between "official clergy" from the Spiritual Boards and informal Islamic elites, their potential government and business support, financial and organizational resources, and the ability to get popular support among the Muslim population. They thus contribute to develop a complex, multifaceted, religious landscape in Russian oil and gas cities.

Conclusion

The intensive migration of Dagestani to the North has several causes that can be formulated in terms of "push-and-pull": lack of jobs and prospects in their region of origin, higher earning potential in Tyumen oblast, and the presence of well-established migrant communities that reduce the cost of moving and offer safety networks for integrating into a new environment. The different migration and integration strategies among North Caucasians are not connected with ethnic identities or with geographic conditions (highlanders versus plains inhabitants), but rather with the socioeconomic circumstances faced by some rural communities, which collectively responded to them through migration.

Everyday discrimination of people from the North Caucasus in KhMAO and YaNAO, as in many other regions of Russia, may be rooted in the long history of the difficult integration of the North Caucasus into the Russian empire and then the Soviet Union, but should also be seen as a product of post-Soviet conditions. On one side, xenophobia is rising among all social groups in Russia, and, on the other, Dagestani migrants bring with them a tradition of distrust toward state structures, especially law-enforcement agencies, and strong habits of working within their own community. However, the second generation of Dagestani living in Western Siberia's oil and gas cities is largely more acculturated to local norms and habits and better integrated socially and culturally.

The dual identities that migrants build for themselves show their high level of adaptability to post-Soviet conditions: mobility is now considered a valorized social norm, through which whole new generations and communities represent themselves and their future. It offers economic prospects as well as upward mobility in terms of accessing better education, healthcare, retirement opportunities, and the possibility of traveling throughout Russia. Whereas Russian authorities try to maintain North Caucasian republics and societies in a kind of isolation from the rest of the country, these mass migrations and the workforce needs of Northern cities help to integrate North Caucasians into the Russian social fabric more than ever, with long-standing impacts, both for the community of origin, in which being a "Northerner" is a sign of social prestige and recognition, and for the receiving community, which is becoming more multicultural and multireligious.

Notes

1 Moscow has experienced a residential and commercial construction boom in the past 15 years. These projects, as well as construction of facilities in Sochi for the Olympic Games and the APEC summit in Vladivostok, turned these regions into large construction sites for two or three years, attracting migrant workers from across the country and from neighboring states.
2 The largest wholesale and retail markets in Moscow. From the late 1990s through the end of the 2000s, a considerable proportion of imported, smuggled, and counterfeit goods passed through here en route to all regions of the Russian Federation (except the Far East Federal Okrug).
3 "Uchastnik vooruzhennogo avtoprobega v Surgute ob"iastnil zachem vzyal pistolety." Ntv.ru. May 6, 2010. Accessed January 24, 2016. www.ntv.ru/novosti/583860/.
4 "V ugolovnom dele po faktu ubiistva surgutskikh chinovnikov, skoree vsego, eshche ne postavlena tochka." SiTV.ru, November 28, 2013. http://sitv.ru/arhiv/news/social/63290/.
5 A national melody and type of dance of many nations of the Caucasus.

6 A doctor in Dagestan would be paid around 12,000 rubles, while for the same job in Surgut he would receive 35,000, including the Northern allowances. In addition, there would be a 10 percent bonus after the first year of work, and 10 percent every six months thereafter, up to 50 percent in KhMAO, and up to 80 percent in YaNAO. These bonuses do not play a significant role, as they are accrued only to the salary at the main workplace and not to the total earnings.

7 The day Stalin deported the Chechens in "Operation Chechevitsa" (lentil) in 1944.

8 Karamahi is a Dargin village in Dagestan, which has about 500 heavy-load vehicles in private ownership, the owners of which are engaged in cargo transportation across the entire country.

9 There are "dachas" near Novaya Fedorovka that spontaneously popped up on vacant land.

10 "Regional'naia vlast' za reformirovanie tiumenskoi ummy." IslamNews, May 12, 2014. www.islamnews.ru/news-145895.html.

Bibliography

Adeeb, Khalid. 2010. *Islam posle kommunizma: religiia i politika v Tsentral'noi Azii.* Moscow: NLO.

Aganbegyan, A.G. et al., eds. 1984. *BAM: stroitel'stvo i khoziaistvennoe osvoenie.* Moscow: Ekonomika.

Aglarov, Mamaikhan. 1988. *Sel'skaia obshchina v Dagestane v XVII–XIX vv.* Moscow, Nauka.

Alanen, Ilkka, Jouko Nikula, Helvi Poder, and Rein Ruutsoo, eds. 2001. *Decollectivisation, Destruction, and Disillusionment: A Community Study in Southern Estonia.* Burlington, VT: Ashgate.

Aleksandrovna, Anna, and Ekaterina Aigina. 2014. "Ethno-Tourism Research in Lovozero, Murmansk Region," SHS Web of Conferences, 4th International Conference on Tourism Research, vol. 12. www.shs-conferences.org/articles/shsconf/abs/2014/09/shsconf_4ictr2014_01036/shsconf_4ictr2014_01036.html.

Aleshkevich, E.S. 2010. "Implementation of Policy Decisions to Develop the Far North: History of the Long-distance Commute Work." In *Biography, Shift-labour, and Socialisation in a Northern Industrial City – The Far North: Particularities of Labour and Human Socialisation,* edited by F. Stammler and G. Eilmsteiner-Saxinger, 109–115. Vienna/Rovaniemi: University of Vienna/Arctic Centre University of Lapland. Accessed January 1, 2016, https://raumforschung.univie.ac.at/fileadmin/user_upload/inst_geograph/BOOK_Biography-ShiftLabour-Socialisation-Russian_North.pdf.

Ananenkov, A.G., G.P. Stavkin, O.P. Andreev, Z.A. Salichov, V.S. Kramar, A.K. Arabskii, V.A. Borovikov, I.A. Orlova, A.N. Fomin, S.V. Okhotnikov, and A.K. Sobakin. 2005. *Sotsial'nye aspekty tekhnicheskogo regulirovaniia vakhtovogo metoda raboty v usloviiakh Krainego Severa.* Moscow: Nedra.

Anderson, David G. 1991. "Turning Hunters into Herders: A Critical Examination of Soviet Development Policy among the Evenki of Southeastern Siberia." *Arctic (Journal of the Arctic Institute of North America)* 44 (1): 12–22.

Anderson, David. 2000. "The Evenks of Central Siberia." In *Endangered Peoples of the Arctic,* edited by M.M. Freeman, 59–74. Westport, CT: Greenwood.

Andreyev, O.P., A.K. Arabskii, V.A. Borovikov, and V.S. Kramar. 2007. "Sistema menedzhmenta vakhtovogo metoda raboty predpriiatiia v usloviiakh Krainego Severa." *Gazovaia promyshlennost'* 9: 28–31.

Andreyev, O.P., A.K. Arabskiy, V.S. Kramar, and A.N. und Silin. 2009. "Sistema menedzhmenta vakhtovogo metoda raboty predpriyatiya v usloviyakh Kraynego Severa." Moskva: Nedra.

Animitsa, E.G. et al. 2010. *Kontseptual'nye podkhody k razrabotke strategii razvitiia monoprofil'nogo goroda*. Ekaterinburg: Izdatel'stvo URGU.

Anisimova, Ia. 2014. "Otgoloski voyny". *Slovo neftianika*. June 27. http://slovon.ru/index .php/sotsium/717-otgoloski-vojny.html.

Argudiaeva, Iu. V. 1988. *Trud i byt molodezhi BAMa: nastoiashchee i budushchee*. Moscow: Mysl'.

Bardileva, Iu. L., and A.K. Portsel. 2014. "Sovremennye demograficheskie vyzovy i ugrozy v Murmanskoi oblasti." In *Sovremennye vyzovy i ugrozy razvitiia v Murmanskoi oblasti: regional'nyi atlas*. Murmansk: Murmansk State Pedagogical University.

Barsukova, Svetlana, and Vadim Radaev. 2012. "Neformal'naia ekonomika v Rossii: kratkii obzor." *Ekonomicheskaia sotsiologiia* 13 (4): 99–11.

Barth, Frederik. 2006. *Etnicheskie gruppy i sotsial'nye granitsi*. Moscow: New Edition.

Beach, Ruth, David Brereton, and David Cliff. 2003. *Workforce Turnover in FIFO Mining Operations in Australia: An Exploratory Study*. Brisbane: Centre for Social Responsibility in Mining, Sustainable Minerals Institute, University of Queensland.

Belkin, E.V. and F.E. Sheregi. 1985. *Formirovanie naseleniia v zone BAMa*. Moscow: Mysl'.

Beneria, Lourdes. 1989. "Subcontracting and Employment Dynamics in Mexico City." In *The Informal Economy: Studies in Advanced and Less Developed Countries*, edited by Alejandro Portes, Manuel Castells, and Lauren A. Benton, 173–188. Baltimore, MD: The John Hopkins University Press.

Berkowitz, Daniel and David N. DeJong. 2001. "Entrepreneurship and Post-Socialist Growth." William Davidson Working Paper No. 406.

Bobrovnikov, Vladimir O. 1997. "Islam i sovetskoe nasledie v kolkhozakh Severo-Zapadnogo Dagestana." *Etnograficheskoe obozrenie* 5: 132–142.

Bobrovnikov Vladimir O. 2001. "Ierarkhiia i vlast' v gornoi Dagestanskoi obchshine", *Rasy i narody* 26: 96–107.

Bogaturov, A.D., A.S. Dundich, and V.G. Korgun. 2011. *International Relations in Central Asia. Events and Documents*. Moscow: Aspect Press.

Bogatyr', Natal'ia V. 2013. "Peredavaia retsepty: kak rasprostraniaiutsia pol'zovatel'skie innovatsii." *Ekonomicheskaia sotsiologiia* 14 (5): 73–103.

Bogdanova, E. 2013. "Kak utopiia stala real'nost'iu. Stroitel'stvo BAMa – samoe shchastlivoe vremia v moei zhizni". In *Topografiia shchast'ia. Etnograficheskie karty moderna*, edited by N.N. Ssorin-Chaikov. Moscow: NLO.

Bolotova A., and F. Stammler. 2010. "How the North Became Home: Attachment to Place Among Industrial Migrants in Murmansk Region." In *Migration in the Circumpolar North: Issues and Contexts*, edited by L. Huskey and Ch. Southcott, 193–220. Edmonton: University of the Arctic, Canadian Circumpolar Institute.

Bondarenko, G.V., E.S. Gareev, and A.N. Kharitonov. 2003. "Vakhtovo-ekspeditsionnaia sistema proizvodstvennoi deiatel'nosti predpriatii TEK na Krainem Severe v kontekste strategicheskikh budushchego razvitiia Rossiiskogo obshchestva." *Gumanitarnye nauki* 14: 252–263.

Boschma, Ron. 2005. "Proximity and Innovation: A Critical Assessment." *Regional Studies* 39 (1): 61–74.

Bourdieu, Pierre. 1977. *Outline of a Theory of Practice*. Cambridge: Cambridge University Press.

Bourdieu, Pierre. 1984. *Distinction: A Social Critique of the Judgement of Taste*. Cambridge, MA: Harvard University Press.

Brandtstädter, Susanne. 2003. "The Moral Economy of Kinship and Property in Southern China." In *The Postsocialist Agrarian Question: Property Relations and the Rural Condition*, edited by Chris Hann, 419–440. Munich: LIT Verlag.

Brezhnev, V.A., ed. 1993. *Baikalo-Amurskaia zheleznodorozhnaia magistral': tekhnicheskii otchet. Letopis' trudovykh otnoshenii na stroitel'stve BAMa. 1967–1989.* Moscow: OAO Korporatsiia Transstroy.

Bublitz, H. 2003. "Diskurs und Habitus." In *"Normalität" im Diskursnetz soziologischer Begriffe*, edited by J. Link, T. Loer, and H. Neuendorff, 151–162. Heidelberg: Synchron Wissenschaftsverlag der Autoren.

"Budushchee sozdaietsia segodnia." 2013. *Rossiiskaia gazeta* (June 28): A16–A17.

Bulaev, V.M. 1998. *Etno-natsional'nye osobennosti formirovaniia naseleniia vostochnogo Zabaikal'ia.* Ulan-Ude: Izdatel'stvo Buriatskogo nauchnogo tsentra RAN, Sibirskoe otdelenie.

Bykov, V.M. 2011. *Formirovanie konkurentosposobnogo personala v usloviiakh vakhtovogo metoda raboty. Na primere neftegazovoi otrasli.* Yaroslavl': Avers Plyus.

Cartwright, A.L. 2001. *The Return of the Peasant. Land Reform in Post-Communist Romania.* Burlington, VT: Ashgate.

Castel, Robert. 2003. *L'insécurité sociale.* Paris: Seuil.

"Chast' Krasnoiarskogo kraia predlagaiut vkliuchit' v sostav Arkticheskoi zony RF." 2013. *Regnum.* August 12. Accessed December 3, 2015, http://regnum.ru/news/1693950. html#ixzz2fr0U6aQr.

Chekmyshev, O.A., and A.D. Yashunskii. 2014. "Izvlechenie i ispol'zovanie dannykh iz elektronnykh sotsial'nykh setei." Moscow: Keldysh Institute of Applied Mathematics RAS.

Chistiakova, Anna. 2015. "Dlinnyi remont za korotkoe leto." *Rossiiskaia gazeta*, March 24. www.rg.ru/2015/03/24/reg-szfo/remont.html.

Colantonio, Andrea. 2007. "Social Sustainability: An Exploratory Analysis of its Definition, Assessment Methods, Metrics, and Tools." Oxford Institute for Sustainable Development, EIBUS Working Paper Series, No. 2007/01.

D'Arcus, Bruce. 2003. "Contested Boundaries: Native Sovereignty and State Power at Wounded Knee, 1973." *Political Geography* 22 (4): 415–437.

Davidova, Sophia, and Kenneth J. Thomson. 2003. *Romanian Agriculture and Transition toward the EU.* Lanham, MD: Lexington Books.

"Demograficheskoe razvitie i etnicheskie protsessy." ND. *Kol'skaia entsyklopediia*, http:// ke.culture.gov-murman.ru/murmanskaya_oblast/5238/.

Didyk, V.V., L.A. Riabova, and E.E. Emel'ianova. 2014. "Monogoroda: sovremennoe polozhenie, kliuchevye problem, i puti ikh resheniia." In *Sovremennye vyzovy i ugrozy razvitiia v Murmanskoi oblasti: regional'nyi atlas.* Murmansk: Murmansk State Pedagogical University.

Dienes, Leslie. 2002. "Reflections on a Geographic Dichotomy: Archipelago Russia." *Eurasian Geography and Economics* 43 (6): 443–458.

Durkheim, Emile. 1960. *The Elementary Forms of Religious Life.* Glencoe, IL: Free Press.

Dybbroe, Susanne. 2008. "Is the Arctic Really Urbanizing?" *Etudes/Inuits/Studies* 32 (1): 13–32.

Efremov, Igor. 2014. "Vozrastnye osobennosti migratsionnykh protsessov na Krainem Severe Rossii," *Demoscope*, http://demoscope.ru/weekly/2014/0581/analit06.php.

Egorov, E.G. 2006. *Sever Rossii: ekonomika, politika, nauka.* Yakutsk: Sakhapoligrafizdat.

Eilmsteiner-Saxinger, Gertrude. 2010. "Multiple Locality and Socially Constructed Spaces among Interregional *Vakhtoviki*: HOME – JOURNEY – ON DUTY."

In *Biography, Shift-labour, and Socialisation in a Northern Industrial City – The Far North: Particularities of Labour and Human Socialisation*, edited by F. Stammler and G. Eilmsteiner-Saxinger, 116–125. Vienna/Rovaniemi: University of Vienna/ Arctic Centre University of Lapland. Accessed January 1, 2016, https://raumforschung .univie.ac.at/fileadmin/user_upload/inst_geograph/BOOK_Biography-ShiftLabour-Socialisation-Russian_North.pdf.

Eilmsteiner-Saxinger, Gertrude. 2011. "'We Feed the Nation': Benefits and Challenges of Simultaneous Use of Resident and Long-distance Commuting Labour in Russia's Northern Hydrocarbon Industry." *Journal of Contemporary Issues in Business and Government* 17 (1): 53–67.

Eilmsteiner-Saxinger, Gertrude. 2013. "Bodenschätze und Menschenschätze – Zur sozialen und materialen Dimension der fossilen Rohstoffe in Nordwest-Sibirien im Kontext des Fernpendelns." In *Wege zum Norden. Wiener Forschungen zu Arktis und Subarktis*, edited by S. Donecker, I. Eberhard, and M. Hirnsperger, 23–43. Vienna: LIT.

Eilmsteiner-Saxinger, Gertrude and E.S. Aleshkevich. 2008. "State and Shift Labor in Western Siberia." Paper presented at the ICASS VI Conference, Nuuk, Greenland, August 26. Estrin, Saul, Ruta Aidis, and Tomasz Mickiewicz. 2007. "Institutions and Entrepreneurship Development in Russia: A Comparative Perspective." William Davidson Institute Working Paper No. 867, February 2007.

Estrin, Saul, Ruta Aidis, and Tomasz Mickiewicz. 2006. "Institutions and Entrepreneurship Development in Russia: A Comparative Perspective." *William Davidson Institute Working Paper Number 867 February 2007.*

Fedorov, P.F. 2009. *Severnyi vektor v rossiiskoi istorii. Tsentr i kol'skoe zapoliar'e v 16–20 vekov.* Murmansk: Murmansk State Pedagogical University.

Fedorov, P.F. 2014. *Kul'turnye landshafty kol'skogo severa v usloviakh urbanizatsii (1931–1991 gg).* Murmansk: Murmansk State Pedagogical University.

Foek, Anton. 2008. "Norilsk Nickel: A Tale of Unbridled Capitalism, Russian Style." CorpWatch. Accessed June 2012, www.corpwatch.org/article.php?id=15215.

Fondahl, Gail. 1998. *Gaining Ground? Evenkis, Land, and Reform in Southeastern Siberia.* Boston: Allyn and Bacon.

Fondahl, Gail and Anna Sirina. 2006. "Rights and Risks: Evenki Concerns Regarding the Proposed Eastern Siberia-Pacific Ocean Pipeline." *Sibirica* 5 (2): 115–118.

Gareyev, E.S., Iu.N. Dorozhkin, G.V. Bondarenko, and I.M. Oreshnikov. 2002. *Sotsial'nyye aspekty perekhoda gradoobrazuiushchego predpriiatiia TEK Severa k vakhtovo-ekspeditsionnomu metodu raboty.* Ufa: UGNTU.

Gerhard, von Ute, Walter Grünzweig, Jürgen Link, and Rolf Parr. 2003. *(Nicht) normale Fahrten. Faszination eines modernen Narrationstyps.* Heidelberg: Synchron Wissenschaftsverlag der Autoren.

Gibson, James L. 2001. "Social Networks, Civil Society, and the Prospects for Consolidating Russia's Democratic Transition." *American Journal of Political Science* 45 (1): 51–69.

Glaser, Barney G., and Anselm L. Strauss. 1967. *The Discovery of Grounded Theory: Strategies for Qualitative Research.* New York: Aldine de Gruyter.

Goode, J. Paul. 2011. *The Decline of Regionalism in Putin's Russia: Boundary Issues.* London and New York: Routledge.

Gorchakov, Rostislav. "Tri novelly ob Igarke." *Memorial.* Accessed November 20, 2015, www.memorial.krsk.ru/memuar/Gorchakov.htm.

Granovetter, Mark S. 1973. "The Strength of Weak Ties." *American Journal of Sociology* 78 (6): 1360–1380.

Granovetter, Mark S. 1985. "Economic Action and Social Structure: The Problem of Embeddedness." *American Journal of Sociology* 91 (3): 481–510.

Gudeman, Stephen. 2008. *Economy's Tensions: The Dialectics of Community and Market.* New York: Berghahn Books.

Habeck, Joachim Otto. 2002. "How to Turn a Reindeer Pasture into an Oil Well and Vice Versa: Transfer of Land, Compensation, and Reclamation in the Komi Republic." In *People and the Land: Pathways to Reform in Post-Soviet Siberia*, edited by Erich Kasten, 125–147. Berlin: Dietrich Reimer Verlag.

Hann, Chris, and Keith Hart. 2011. *Economic Anthropology: History, Etnography, Critique.* Cambridge, UK: Polity.

Hann, Christopher M. 2000. "The Tragedy of the Privates? Postsocialist Property Relations in Anthropological Perspective." Max Planck Institute for Social Anthropology Working Paper No. 2, 1–23.

Heininen, Lassi, Alexander Sergunin, and Gleb Yarovoy. 2013. "New Russian Arctic Doctrine: From Idealism to Realism?" Valdai Discussion Club. July 15. Accessed September 1, 2014, http://valdaiclub.com/russia_and_the_world/60220.html.

Heleniak, Timothy. 1999. "Out-migration and Depopulation of the Russian North during the 1990s." *Post-Soviet Geography and Economics* 40 (3): 281–304.

Heleniak, Timothy. 2009a. "Growth Poles and Ghost Towns in the Russian Far North." In *Russia and the North*, edited by Elana Wilson Rowe, 129–164. Ottawa: University of Ottawa Press.

Heleniak, Timothy. 2009b. "The Role of Attachment to Place in Migration Decisions of the Population of the Russian North." *Polar Geography* 32 (1–2): 31–60.

Heleniak, Timothy. 2010. "Population Change in the Periphery: Changing Migration Patterns in the Russian North." *Sibirica: Interdisciplinary Journal of Siberian Studies* 9 (3): 9–40.

Henry, Laura. 2009. "Thinking Globally, Limited Locally: The Russian Environmental Movement and Sustainable Development." In *Environmental Justice and Sustainability in the Former Soviet Union*, edited by Julian Agyeman and Yelena Ogneva-Himmelberger, 47–70. Cambridge, MA: The MIT Press.

Hill, Fiona, and Clifford Gaddy. 2003. *The Siberian Curse: How Communist Planners Left Russia out in the Cold.* Washington DC: Brookings Institution Press.

Hobart, C.W. 1979. "Commuting Work in the Canadian North: Some Effects on Native People." In *Proceedings of Conference on Commuting and Northern Development*, edited by M. Mougeot, 1–38. Saskatoon: Institute for Northern Studies, University of Saskatchewan.

Humphrey, Caroline. 2002. *The Unmaking of Soviet Life: Everyday Economies after Socialism.* Ithaca, NY: Cornell University Press.

Huskey, Lee. 2010. "Globalization and the Economies of the North." In *Globalization and the Circumpolar North*, edited by Lassi Heininen and Chris Southcott, 57–90. Fairbanks: University of Alaska Press.

Huskey, Lee, Matthew Berman, and Alexandra Hill. 2004. "Leaving Home, Returning Home: Migration as a Labor Market Choice for Alaska Natives." *Annals of Regional Science* 38: 75–92.

Ivanishcheva, O.N. 2014. "Saami kol'skogo severa. Problemy sokhraneniia iazyka, kul'tury i traditsionnogo khozaistva." In *Sovremennye vyzovy i ugrozy razvitiia v Murmanskoi oblasti: regional'nyi atlas*, 246–251.

Josephson, Paul R. 2014. *The Conquest of the Russian Arctic.* Cambridge, MA: Harvard University Press.

"Kadry dlia Severa. Noril'skii Nikel' preodolevaet defitsit kvalifitsirovannogo personala." 2015. *Izvestiia*, August 29, http://izvestia.ru/news/519193.

Kapustina, Ekaterina L. 2012. "Vybory v sel'skom Dagestane: politicheskoe sobytie kak element sotsial'noi zhizni." In *Obshchestvo kak ob"ekt i sub"ekt vlasti. Ocherki politicheskoi antropologii Kavkaza. Peterburzhskoe vostokovedenie*, 32–61.

Kapustina, Ekaterina L. 2013. "Trudovaia migratsiia iz sel'skogo Dagestana kak khozaistvennaia praktika i sotsiokul'turnoe iavlenie." Unpublished paper. http://kunstkamera.ru/files/doc/autoreferat_kapustina_e_l.pdf.

Kapustina, Ekaterina L. 2014. "Sobstvennost' na Sever: migrant iz Dagestana i osvoenie gorodskogo prostranstva v Zapadnoi Sibiri (na primere situatsii v g. Surgut), *Journal of Sociology and Social Anthropology* 5: 12–25.

Karachurina, L. 2012. "Urbanizatsiia 'po otchety' i 'po fakty'." *Demoscope Weekly*, August 20–September 2. http://demoscope.ru/weekly/2012/0519/tema06.php#_FNR_53.

Karpov, Iurii Iu. 2007a. "Gornoe dagestanskoe selenie: ot traditsionnogo dzhamaata k nyneshnemu sotsial'nomu bliku." In *Severnyi Kavkaz: traditsionnoe sel'skoe soobshchestvo. Sotsial'nye roli, obshchestvennoe mnenie, vlastnye polnomoch'ia. Nauka*, 5–72.

Karpov, Iuri Iu. 2007b. *Vzgliad na gortsev, vzgliad s gor: mirovozzrencheskie aspekty kul'tury i sotsial'nyi opyt gortsev Dagestana*. St. Petersburg: Peterburgskoe vostokovedenie.

Karpov, Iuri Iu., and Ekaterina L. Kapustina. 2011. *Gor gortsy posle gor. Migratsionnye protsessy v Dagestane v XX – nachale XXI veka: ikh sotsial'nye i etnokul'turnye posledstviia i perspektivy*. St. Petersburg: Peterburgskoe vostokovedenie.

Kishigami, Nobuhiro, and Molly Lee. 2008. "Urban Inuits." *Etudes/Inuit/Studies* 32 (1): 9–11.

Kliueva, V.P. 2001. "Bukharskie obshchiny v Sibiri (konets XVI – nachalo XIX vv.)." In *Problemy ekonomicheskoi i sotsial'no-politicheskoi istorii dorevolutsionnoi Rossii*. Tyumen, 77–85.

Köllner, Tobias. 2013. *Practising Without Belonging? Entrepreneurship, Morality, and Religion in Contemporary Russia*. Munich: LIT Verlag.

"Komu pomeshal imam mecheti Novogo Urengoia." 2010. *Muslim-info.ru*, December. http://muslim-info.ru/news.php?id=127.

Kontysheva, O. 2007. "Vladimir Kokorin – The Oilman." *Kopeika*. May 12. http://pressa.irk.ru/kopeika/2007/48/010001.html.

Kozlynskaya, N.M. 2009. "Osobennosti regulirovaniia truda lits, rabotaiushchikh v rayonakh Krainego Severa i priravnennykh k nim mestnostiakh." PhD diss., Moscow, Russian Academy. Accessed January 1, 2016, www.dissercat.com/content/osobennosti-regulirovaniya-truda-lits-rabotayushc‚hikh-v-raionakh-krainego-severa-i-priravnen.

Krivoy, V.I., ed. 1989. *Vakhtovyy metod: pravovyye voprosy*. Moscow: Iuridicheskaia literatura.

Kronenberg, Tobias. 2004. "The Curse of Natural Resources in the Transition Economies." *Economics of Transition* 12 (3): 399–426.

Kuznetsov, Andrei, and Olga Kuznetsova. 2005. "Business Culture in Modern Russia: Deterrents and Influences." *Problems and Perspectives in Management* 2 (2): 25–31.

Larsen, Joan Nymand and Gail Fondahl, eds. 2015. *Arctic Human Development Report: Regional Processes and Global Linkages*. Copenhagen: Nordic Ministry.

Laruelle, Marlene. 2013. Unpublished fieldwork notes, Norilsk.

Laruelle, Marlene. 2014. *Russia's Arctic Strategies and the Future of the Far North*. Armonk, NY: M.E. Sharpe.

Laruelle, Marlene. 2016. "Assessing Social Sustainability. Immigration to Russia's Arctic Cities." In *Sustaining Russia's Arctic Cities*, edited by Robert Orttung. Oxford, New York: Bergham.

Ledeneva, Alena V. 1998. *Russia's Economy of Favours. Blat, Networking and Informal Exchange*. Cambridge: Cambridge University Press.

Ledeneva, Alena. 2000. "Shadow Barter: Economic Necessity or Economic Crime?" In *The Vanishing Rouble: Barter Networks and Non-Monetary Transactions in Post-Soviet Societies*, edited by Paul Seabright, 298–317. New York: Cambridge University Press.

Ledeneva, Alena V. 2006. *How Russia Really Works. The Informal Practices that Shaped Post-Soviet Politics and Business*. Ithaca and London: Cornell University Press.

Link, J. 1997. *Versuch über den Normalismus: Wie Normalität produziert wird*. Opladen: Westdeutscher Verlag.

Lobo, Susan. 2001. "Is Urban a Person or Place? Characteristics of Urban Indian Country." In *The American Indians and the Urban Experience*, edited by Susan Lobo and Kurt Peters, 73–84. Lanham, MD: Altamira Press.

Lobo, Susan, and Evelin Peters. 2001. *American Indians and the Urban Experience*. Lanham, MD: Altamira Press.

Magomedov, Khabib G., and Denis V. Sokolov. 2011. "Urbanizatsiia i razvitie gorodov na Severnom Kavkaze." In *Severnyi Kavkaz: modernizatsionnyi vyzov*, edited by I.V. Starodubrovskaya et al., 235–265. Moscow: Delo.

Malashenko, Aleksei V. 1998. *Islamskoe vozrozhdenie v sovremennoi Rossii*. Moscow: Carnegie Center, 78–79.

Marshall, Alfred. 1920. *Principles of Economics*. London: Macmillan.

Martynov, V.A. 2010. "The Organization of the Long-Distance Commute Work in the Far North and its Legal Bases." In *Biography, Shift-labour, and Socialisation in a Northern Industrial City – The Far North: Particularities of Labour and Human Socialisation*, edited by F. Stammler and G. Eilmsteiner-Saxinger, 85–88. Vienna/Rovaniemi: University of Vienna/Arctic Centre University of Lapland. Accessed January 1, 2016, https://raumforschung.univie.ac.at/fileadmin/user_upload/inst_geograph/BOOK_Biography-ShiftLabour-Socialisation-Russian_North.pdf.

Martynov, V.G., and A.A. Moskalenko. 2008. *Kadrovaia politika kak instrument v sisteme sotsial'noi otvetstvennosti biznesa*. Moscow: MAKS Press.

McCannon, John. 1998. *Red Arctic: Polar Exploration and the Myth of the North in the Soviet Union, 1932–1939*. New York: Oxford University Press.

McCarthy, Danel J., Sheila Puffer, and Stanislav V. Shekshnia. 1993. "The Resurgence of an Entrepreneurial Class in Russia." *Journal of Management Inquiry* 2 (2): 125–137.

"Medrese 'Belem,' putevka v zhizn' dlia musul'man Tyumeni." 2012. *Islamrf.ru*, September 29. www.islamrf.ru/news/events/russia/24207.

Metzo, Katherine. 2009. "Civil Society and the Debate over Pipelines in Tunka National Park, Russia." In *Environmental Justice and Sustainability in the Former Soviet Union*, edited by Julian Agyeman and Yelena Ogneva-Himmelberger, 119–140. Cambridge, MA: The MIT Press.

Mikhailov, Andre. 2013. "Murmansk: goro zimy v otluchke i radugi." *Pravda*, August 7.

Mikhailov, E.I. 2004. *Migratsionnye protsessy v istorii formirovaniia naseleniia evropeiskogo severa rossii v XX veke*. Kandidat diss., Murmansk Scientific Institute.

Ministry of Labor and Employment. 2013. "Analysis of Income of Workers in Irkutsk Oblast in January–June 2013." Accessed December 28, 2013, www.irkzan.ru/qa/1498.html.

Moe, Arild, and Valery Kryukov. 2010. "Oil Exploration in Russia: Prospects for Reforming a Crucial Sector." *Eurasian Geography and Economics* 51 (3): 312–329.

Mote, Victor. 2003. "Stalin's Railway to Nowhere: 'The Dead Road' (1947–1953)." *Sibirica* 3 (1): 8–63.

MRD (Federal Ministry of Regional Development). 2013. *Concept Paper on the Draft Federal Law "On the Arctic zone of the Russian Federation."* Moscow, Russian Federation. Accessed October 6, 2013, www.minregion.ru/upload/documents/2013/04/080413/080413_p_k.doc.

Murmansk oblast. 2014a. "Murmanskaia oblast' otmechena Minregionom Rossii v chisel naibolee aktivnykh uchastnikov razrabotki osnov gosudarstvennoi politiki v Arktike." March 6. http://new.gov-murman.ru/info/news/884/?sphrase_id=71741.

Murmansk oblast. 2014b. "Gubernator Murmanskoi oblasti Marina Kovtun prizyvaet vse arkticheskie regiony k konsolidatsii." November 18. http://new.gov-murman.ru/info/news/48539/?sphrase_id=71739.

Nee, Victor. 1991. "Peasant Entrepreneurs in China's 'Second Economy': An Institutional Analysis." *Economic Development and Cultural Change* 39 (2): 293–310.

Norilsk. 2013. "Proposals on the draft Federal Law 'On the Arctic zone of the Russian Federation.'" Report presented at 33rd Congress of the Union of Cities in the Arctic and the High North, Norilsk, Krasnoyarsk krai, Russian Federation, July 22.

Nuykina, Elena. 2011. *Resettlement from the Russian North: An Analysis of State-induced Relocation Policy.* Rovaniemi, Finland: Arctic Centre, University of Lapland.

"O musul'manskom sovte, Maugli i bidga." 2011. *IslamNews*, June 30. Accessed January 23, 2016, www.islamnews.ru/news-66551.html.

Öfner, Elisabeth. 2014. "Russia's Long-distance Commuters in the Oil and Gas Industry: Social Mobility and Current Developments: An Ethnographic Perspective from the Republic of Bashkortostan." *Journal of Rural and Community Development* 9 (1): 41–56.

Öhman, Marianne, and Urban Lindgren. 2003. "Who Is the Long-Distance Commuter? Patterns and Driving Forces in Sweden." *Cybergeo: European Journal of Geography* 243. Accessed January 1, 2016, http://cybergeo.revues.org/4118.

Orttung, Robert. 2006. "Business-State Relations in Russia." *Russian Analytical Digest,* Issue 8. Center for Security Studies (CSS), ETH Zurich; Research Centre for East European Studies, University of Bremen.

Ostreng, Willy. 2010. "The Russian Federation's Arctic Policy, CHNL. ARCTIS Knowledge Hub." Accessed December 5, 2015, www.arctis-search.com/The+Russian+Federation's+Arctic+Policy.

Oswald, Ingrid. 2012. "The Industrialized Village. To Transformation of the Rural Way of Life in Post-Socialist Societies." In *Away from the City: Life in Post-Soviet Villages,* edited by Elena Bogdanova and Olga Brednikova, 8–27. St. Petersburg: Aletheia.

Overland, Indra. 2009. "Indigenous Rights in the Russian North." In *Russia and the North,* edited by Elana Wilson Rowe, 165–186. Ottawa: University of Ottawa Press.

Pacific Environment. 2013. "Siberia-Pacific pipeline." Pacific Environment. Accessed May 28, http://pacificenvironment.org/siberia-pacific-pipeline.

Parry, Jonathan, and Maurice Bloch. 1989. *Money and the Morality of Exchange.* Cambridge: Cambridge University Press.

Pashin, S.T. 2004. "My znali – Iamburg budet: 20 let Yamburggazdobycha." In *Gazovye okeany Iamburga,* edited by A. Belov, K. Boris, and V. Bashuk, 62–66. Moscow: Kniga-Penta.

Peng, Mike W. 2001. "How Entrepreneurs Create Wealth in Transition Economies." *Academy of Management Perspective* 15 (1): 95–108.

Perepis'. 2010. Russian National Census of 2010. www.perepis-2010.ru/results_of_the_census/results-inform.php.

Petrov, Andrey. 2008. "Lost Generations? Indigenous Population of the Russian North in the Post-Soviet Era." *Canadian Studies in Population* 35 (2): 269–290.

Petrov, Nikolay. 2000. "Federalizm po-rossiiski." *Pro et Contra* 5 (1): 7–33.

Petrov, Nikolay. 2004. "Regional Models of Democratic Development." In *Between Dictatorship and Democracy: Russian Post-Communist Political Reform*, edited by Michael McFaul and Nikolay Petrov, 239–267. Washington, DC: Carnegie Endowment for International Peace.

Petrov, Nikolay. 2005. "Democracy in Regions of Russia," *Carnegie Moscow Center Briefing* 7 (9): 9.

Petrov, Nikolay and Michael McFaul, eds. 1998. *Political Almanac of Russia, 1997.* Moscow: Carnegie Moscow Center/Carnegie Endowment for International Peace.

Petrov, Nikolay and Alexei Titkov. 2013. "Reiting demokratichnosti Moskovskogo Tsentra Carnegie: 10 let v stroiu," *Carnegie Moscow Center Working Paper* (December).

Petrova, Natalya. 2014. "Gazprom Neft Moves to Optimize Prirazlomnoye Oil Field." *Offshore Magazine*, November 12. www.offshore-mag.com/articles/print/volume-74/issue-11/arctic/gazprom-neft-moves-to-optimize-prirazlomnoye-oil-field.html.

Pickles, John and Adrian Smith, eds. 1998. *Theorising Transition: The Political Economy of Post-Communist Transformation.* London: Routledge.

Pissarides, Francesca, Miroslav Singer, and Jan Svejnar. 2000. "Objectives and Constraints of Entrepreneurs: Evidence from Small and Medium Size Enterprises in Russia and Bulgaria." William Davidson Institute Working Paper No. 346, October 2000.

Polianskaia, Ekaterina. 2013. "Arkticheskii zakonoprekt na finishnoi priamoi." *Pravda*, March 14. Accessed November 20, 2015, www.pravda.ru/society/how/14-03-2013/1148406-hyhn-0/.

Polyakov, O.A. 2013. "Perspektivy osveneiya prirodnykh resursov i razvitiya promyshlennosti rayonov severa i yugo-vostoka Zabaykal'skogo kraia," Presentation by the Vice Minister of Natural Resources and Environment of Zabaikal'skii krai. Chita, August 19.

Povoroznyuk, O.A. 2011. *Zabaikal'skie Evenki: sotsial'nye, ekonomicheskie i kul'turnye transformatsii v XX–XXI vekakh.* Moscow: Institute of Ethnology and Anthropology Press.

Povoroznyuk, O.A. 2014. "Aborigeny, bamovtsy i priezzhie: sotsial'nye otnosheniia na severe Zabaikal'ia." In *Sibirskii sbornik 4. Grani sotsial'nogo: Antropologicheskie perspektivy issledovaniia sotsial'nykh otnoshenii i kul'tury*, edited by V.N. Davydov and D.V. Arzyutov, 38–55. St. Petersburg: MAE RAN.

"Proekt gosprogrammy razvitiia Arktiki vnesut v kabmin RF do noiabria." 2013. *RIA-Novosti.* October 18. Accessed December 7, 2015, http://ria.ru/economy/20131018/970989750.html#ixzz2irqQQQjS.

Puffer, Sheila M., Daniel J. McCarthy, and Max Boisot. 2010. "Entrepreneurship in Russia and China: The Impact of Formal Institutional Voids." *Entrepreneurship Theory and Practice* 34 (3): 441–467.

Putsykina, Anna. 2011. "What Lies Beneath the 'Sleeping Land?'" *Mining Journal.* Special Supplement: May. Accessed August 25, 2014, www.miningjournal.com/supplements/mj-siberia-supplement-0511.

Rabinovich, M.G. 1983. "To the Definition of the 'City' (For the Ethnographic Study)." *Soviet Ethnography* 3: 19–24.

RAO/CIS. 2015. Press Center Note May 14. RAO/CIS, www.rao-offshore.com/ index.php?option=com_content&view=article&id=679%3A2015-05-14-13-06-33&catid=5%3Anews&Itemid=54.

Rasmussen, R.O., 2011, "Megatrends: Proceedings from the First International Conference on Urbanisation in the Arctic." Nordic Council of Ministers, NordRegio. Nordregio Working Paper 2011: 527.

Razumova, I.A. 2004. "Severnyi 'migratsionnyi tekts' postsovetskoi Rossii," *Etnokul'turnye protsessy na Kol'skom Severe*. Apatity.

Revich, B.A., T.L. Kharkova, E.A. Kvasha, D.D. Bogoyavlenskii, A.G. Korovkin, and I.B. Korolev. 2014. "Sociodemographic Limitations of the Sustainable Development of Murmansk Region." *Studies on Russian Economic Development* 25 (2): 201–206.

Riabova, Larisa. 2010. "Community Viability and Well-Being in the Circumpolar North." In *Globalization and the Circumpolar North*, edited by Lassi Heininen and Chris Southcott, chap. 5. Fairbanks: University of Alaska Press.

Robbins, Joel. 2007. "Between Reproduction and Freedom: Morality, Value, and Radical Cultural Change." *Ethnos* 72 (3): 293–314.

Rolshoven, Johanna. 2006. "Woanders daheim. Kulturwissenschaftliche Ansätze zur multilokalen Lebensweise in der Spätmoderne." In *Mobility and Social Change. Mobilität und Sozialer Wandel. A Choice of Texts. Eine Textauswahl. 2001–2006*, edited by Johanna Rolshoven, 28–39. Accessed January 1, 2016, www.uni-graz.at/johanna. rolshoven/jr_textauswahl.pdf.

Rolshoven, Johanna. 2008. "The Temptations of the Provisional: Multilocality as a Way of Life." *Ethnologia Europaea* 37 (1–2): 17–25.

Rolshoven, Johanna. 2009. "Kultur-Bewegungen: Multilokalität als Lebensweise in der Spätmoderne." *Österreichische Zeitschrift für Volkskunde* 63 (112): 285–303.

Rolshoven, Johanna. 2011. "Das Figurativ der Vagabondage: eine Einleitung." In *Vagabunden und Vagabondage: Eine Exploration in bewegliche Lebenswelten*, edited by Johanna Rolshoven and M. Maierhofer, Graz: Institut für Volkskunde und Kulturanthropologie: 8–17.

Rozhanskii, M.Ia., ed. 2002. *Baikal'skaia Sibir'. Fragmenty sotsio-kul'turnoi karty: al'manakh-issledovanie*. Irkutsk: Irkutsk State University.

Rozhansky, M.J. 2002. "Introduction." In *The Baikal Siberia: Fragments of Socio-Cultural Map*. Irkutsk: Irkutsk State University.

Russia, Government of. 2008. "Fundamentals of State Policy of the Russian Federation in the Arctic up to 2020 and Beyond." [Osnovy gosudarstvennoi politiki Rossiiskoi Federatsii v Arktike na period do 2020 goda i dal'neishuiu perspektivu]. Adopted by President D. Medvedev, September 2008, released March 27, 2009, the Security Council of the Russian Federation. Moscow, Russian Federation.

Russia, Government of. 2012. "State Programme: North Caucasus Federal District Development to 2025." Accessed November 20, 2015, http://government.ru/en/ docs/7303.

Russia, Government of. 2013a. "Executive Order: On the Russian Federation's Land Areas in the Arctic Zone." May 2014 Executive Order.

Russia, Government of. 2013b. "Development Strategy of the Russian Arctic and the maintenance of national security for the period up to 2020." English Text. Accessed September 2, 2014, www.iecca.ru/en/legislation/strategies/item/99-the-development-strategy-of-the-arctic-zone-of-the-russian-federation.

Russia, Government of. 2014. "The Socioeconomic Development of the Arctic Zone of the Russian Federation for the Period up to 2020." Accessed August 2, 2014, www .minregion.ru/uploads/attachment/a2823718-1d66-401f-bf15-cbc65403e12f.pdf.

"S"ezd Soiuza gorodov Zapoliar'ia i Krainego Severa: zakonproekt 'Ob Arkticheskoi zone RF' ne otvechaet potrebnostiam Severa i ne uchityvaet ego spetsifiky." 2013. *B-Port.com*, July 26. Accessed November 20, 2015, www.b-port.com/news/item/110060.html.

Sakaeva, Maria M. 2012. "Parlament kak 'okno vozmozhnostei': issledovanie povedeniia predprinimatelei s deputatskim mandatom v khode realizatsii rynochnykh interesov." *Ekonomicheskaia sotsiologiia* 13 (3): 96–122.

Saxinger, Gertrude. 2015. "'To You, to Us, to Oil and Gas' – The Symbolic and Socio-economic Attachment of the Workforce to Oil, Gas, and Its Spaces of Extraction in the Yamal-Nenets and Khanty-Mansi Autonomous Districts in Russia." *Fennia- International Journal of Geography* 193(1): 83–98. doi: 10.11143/45209.

Saxinger, Gertrude. 2016a. "Lured by Oil and Gas: Labour Mobility, Multi-locality and Negotiating Normality and Extreme in the Russian Far North." *The Extractive Industries and Society Journal* (forthcoming). doi:10.1016/j.exis.2015.12.002.

Saxinger, Gertrude. 2016b. *Unterwegs. Mobiles Leben in der Erdgas – und Erdölindustrie in Russlands Arktis*. Vienna: Böhlau.

Saxinger, Gertrude, and Elena Nuykina. 2015. "Vakhtoviki i seks-industriia: mify, osvedomlonnost' i deistvie." In *Psikhologiia ekstremal'nykh professii: materialy Vserossiiskoi nauchno-prakticheskoi konferentsii 18–19 dekabria 2014*, edited by A.Ya. Korneyeva, 157–162. Arkhangelsk: Kira.

Saxinger, Gertrude, Elena Nuykina, and Elisabeth Öfner. 2015. "Mobil'nost' i migrat-siia iz Respubliki Bashkortostan v regiony Krainego Severa Rossii." In *Trud, zania-tost' i chelovecheskoe razvitie*, edited by R.M. Valiachmetova, N.M. Baimurzinoi, and N.M. Lavrenyuk, 225–226. Ufa: Vostochnaia pechat'.

Saxinger, Gertrude, Elena Nuykina, and Elisabeth Öfner. 2016. "The Russian North Connected – The Role of Long-Distance Commute Work for Regional Integration." In *Sustaining Russia's Arctic Cities: Resource Politics, Migration, and Climate Change*, edited by Robert Orttung. London: Berghan.

Saxinger, G., E. Öfner, E. Shakirova, M. Ivanova, M. Iakovlev, and E. Gareev. 2014. "'Ia gotov!': Novoe pokolenie mobil'nykh kadrov v rossiiskoi neftegazovoi promyshlen-nosti." *Sibirskie istoricheskie issledovaniia* 3: 73–103.

Schweitzer, Peter P. and Patty A. Gray. 2000. "The Chukchi and Siberian Yupiit of the Russian Far East." In *Endangered Peoples of the Arctic*, edited by Milton M.R. Freeman, 17–38. Westport, CT: Greenwood.

Seabright, Paul, ed. 2000. *The Vanishing Rouble: Barter Networks and Non-Monetary Transactions in Post-Soviet Societies*. Cambridge: Cambridge University Press.

"Sel'skie etiudi k gorodskomu peizazhu: transformatsiia gorodskogo prostranstva v kontekste migratsionnikh protsessov v Dagestane i sud'ba sel'skikh zemliachestv v Makhachkale nachala XXI veka." 2013. *Compendium of the Museum of Anthropology and Ethnography*, Russian Academy of Sciences, 111–175.

"Seven Regions to be Included in Russia's Arctic Zone: Ministry of Regional Develop-ment." 2013. *RIA-Novosti*. October 22. Accessed September 2, 2014, http://arctic.ru/news/2013/10/seven-regions-be-included-russia%E2%80%99s-arctic-zone-ministry-regional-development.

Sirina, Anna. 2009. "Oil and Gas Development in Russia and Northern Indigenous Peoples." In *Russia and the North*, edited by Elana Wilson Rowe, 187–202. Ottawa: University of Ottawa Press, 2009.

Sirina, Anna N. 2012. "Irkutskaia oblast'." In *Sever i severiane. Sovremennoe polozhenie korennykh malochislennykh narodov Severa, Sibiri i Dal'nego Vostoka Rossii*, 121–135. Moscow: Institute of Ethnography and Anthropology of Russian Academy of Sciences.

Slabukha, A. 2010. "Zodchie Noril'laga (gruppovoi portret – iz nekotorykh tsifr stastistiki)." In *International Conference Monumentalita & Modernita: Architecture and Art in Italy, Germany, and Russia in the "Totalitarian" Period*. www.kapitel-spb.ru/index.php/component/content/article/12-konferent/48-slabuha.

Soboleva, V.N., and S.M. Melnikov. 1999. "Migratsionnye protsessy v Magadanskoi oblasti." *Sotsiologicheskoe issledovanie* 59 (11).

Sokolov, Denis V. 2011. *Severnyi Kavkaz: modernizatsionnyi vyzov*, edited by I.V. Starodubrovskaya et al., 235–265. Moscow: Delo.

Sokolov, Denis V. 2012a. "Dzhamaat protiv kolkhoza." In *Fronty: dvatsat' let posle kolkhoza*. Moscow: RAMCOM.

Sokolov, Denis V. 2012b. "Zatoplennyi mir koisubulintsev: elektrichestvo v obmen na abrikosy." In *Obshchestvo kak ob"ekt i sub"ekt vlasti. Ocherki politicheskoi antopologii Kavkaza*. St. Petersburg: Peterburgskoe vostokovedenie, 61–91.

Sokolov, Denis V. 2013. "Etnichnost' i nasilie na Severnom Kavkaze." In *Severnyi Kavkaz: modernizatsionnyi vyzov*, edited by I.V. Starodubrovskaia et al. 78–96. Moscow: Delo.

Sokolov, Denis V. 2014a. "Islam protiv global'nogo rinka." *Kavpolit*, July 11. http://kavpolit.com/articles/islam_protiv_globalnogo_rynka-11073/.

Sokolov, Denis V. 2014b. "Rol' 'sovesti' v neftegazovoi politike." *Kavpolit*, June 20. http://kavpolit.com/articles/rol_sovesti_v_neftegazovoj_politike-7452/.

Sokolovskii, S.V. 2012. "Sovremennyi etnogenez ili politika identichnosti? Ob ideologii naturalisatsii v sovremennykh sotsial'nykh naukakh." *Etnograficheskoe obozrenie* 2: 77–83.

Spies, Mattias. 2009. *Oil Extraction in Extreme Remoteness: The Organisation of Work and Long-Distance Commuting in Russia's Northern Resource Peripheries*. Joensu: Yhteiskuntatieteellisiä julkaisuja, no. 99. Accessed January 1, 2016, http://epublications.uef.fi/pub/urn_isbn_978-952-219-291-2/urn_isbn_978-952-219-291-2.pdf.

Staalesen, Atle. 2013. "Murmansk Fights Moscow over Arctic Territories." *Barents Observer*, August 5. http://barentsobserver.com/en/arctic/2013/08/murmansk-fights-moscow-over-arctic-territories-05-08.

Stammler, F. 2010. "The City Became the Homeland for its Inhabitants, but Nobody Is Planning to Die Here: Anthropological Reflections on Human Communities in the Northern City." In *Biography, Shift-Labour, and Socialisation in a Northern Industrial City – The Far North: Particularities of Labour and Human Socialisation*. Proceedings of the International Conference in Novy Urengoy, Russia, December 4th–6th, 2008: 41–42. http://arcticcentre.ulapland.fi/docs/NURbook_2ed_100421_final.pdf.

Stark, David. 1992. "Path Dependence and Privatization Strategies in East Central Europe." *East European Politics and Societies* 1 (1): 17–54.

Stark, David. 1996. "Recombined Property in East European Capitalism." *American Journal of Sociology* 101 (4): 993–1027.

State Statistics Bureau of Russian Federation. 2013. "Estestvennoe dvizhenie naseleniia Rossiiskoi Federatsii: rodivshiesia, umershie i estestvennyi prirost naseleniia po sub"ektam Rossiiskoi Federatsii – 2013 g." www.gks.ru/bgd/regl/b13_106/IssWWW.exe/Stg//%3Cextid%3E/%3Cstoragepath%3E::%7C04/tab02.xls.

Stepanova, N.A., and R.R. Nogovitsyn. 2011. *Maloe predprinimatel'stvo severnogo regiona. V usloviakh perekhoda na innovatsionnyi put' razvitiia*. Yakustk: Sfera.

Storey, Keith. 2001. "Fly-in/Fly-out and Fly-over: Mining and Regional Development in Western Australia." *Australian Geographer* 32 (2): 133–148.

Strelnikova, D. 2013. "Zona osobogo vnimaniia." *Zapoliarnaia pravda* No. 106, July 25. Accessed December 5, 2015, http://gazetazp.ru/2013/106/1.

Tarasov, M.E., E.G. Egorov, and G.P. Kulakovskii. 2013. *Vliianie tenevoi ekonomiki na ekonomicheskuiu bezopasnost' regiona*. Yakutsk: Severo-Vostochnyi federal'nyi universitet.

Territorial Body of Russian Bureau of Statistics for KhMAO. 2010. http://khmstat.gks.ru/wps/wcm/connect/rosstat_ts/khmstat/resources/609b09004f19d755aa2bba149d0ea7d8/pub-04-04_%D0%A2%D0%B5%D1%80%D1%80%3D71800+-3.pdf.

"They Were the First." 2007. *Oil and Capital*.

Tirole, Jean. 1996. "A Theory of Collective Reputations (with Applications to the Persistence of Corruption and to Firm Quality)." *Review of Economic Studies* 63 (1): 1–22. www.jstor.org/stable/2298112.

Titov, V., M. Rozhanskii, and Y. Ielokhina. 2007. "Fenomen grazhdanskogo protesta: opyt i uroki Baikal'skogo dvizheniia." *Obshchestvo*, January 12. Accessed May 12, 2012, http://newsbabr.com/?IDE=35255.

"Tiumenskie musul'mane razobralis' v mecheti po poniatiam." 2012. *Islam News*, August 27. www.islamnews.ru/news-135958.html.

Tkachev, Alexei, and Lars Kolvereid. 1999. "Self-Employment Intentions among Russian Students." *Entrepreneurship & Regional Development: An International Journal of Comparative Sociology* 1 (1): 269–280.

Tompson, W. 2006. "Frozen Venezuela? The Resource Curse and Russian Politics." In *Russia's Oil and Natural Gas: Bonanza or Curse?*, edited by M. Elman, 189–212. New York: Anthem Press.

"Too expensive: Gazprom puts Shtokman on hold." 2012. *Russia Today*, August 29. www.rt.com/business/shtokman-gas-gazprom-857/.

Torre, Andre. 2008. "On the Role Played by Temporary Geographical Proximity in Knowledge Transfer." *Regional Studies* 42 (6): 869–889.

Torre, Andre. 2011. "The Role of Proximity during Long-Distance Collaborative Projects: Temporary Geographical Proximity Helps." *International Journal of Foresight and Innovation Policy* 7 (1/2/3): 213–230.

Torre, Andre, and Jean-Pierre Gilly. 1999. "On the Analytical Dimension of Proximity Dynamics." *Regional Studies* 34 (2): 169–180.

Torre, Andre, and Frederic Wallet. 2014. "The Role of Proximity Relations in Regional and Territorial Development Processes." Paper presented at the 54th ERSA Congress, St. Petersburg, August 26–29, 2014. Accessed November 20, 2015, http://EconPapers.repec.org/RePEc:wiw:wiwrsa:ersa10p792.

Torsello, Davide. 2003. "Trust and Property in Historical Perspective: Villagers and the Agricultural Cooperative in Királyfa, Southern Slovakia." In *The Postsocialist Agrarian Question: Property Relations and the Rural Condition*, edited by Chris Hann, 93–116. Munich: LIT Verlag.

Tsukerman, V.A., and E.S. Goriachevskaia. 2014. "Osnovnye vyzovy i ogranicheniia innovatsionno-tekhnologicheskogo razvitiia promyshlennosti Murmanskoi oblasti." In *Sovremennye vyzovy i ugrozy razvitiia v Murmanskoi oblasti: regional'nyi atlas*. Murmansk: Murmansk State Pedagogical University.

Turov, M.G. 2008. *Evenki. The Core Issues of Ethno-genesis and Ethnic History*. Irkutsk: "Amtera."

UN Habitat. 2009. "Global'nyi doklad o naselennykh punktakh. Planirovaniie ustoichivykh gorodov: napravleniia strategii." Programma Organizatsii Obyedinennykh Natsii po naselennym punktam. Accessed January 13, 2014, http://unhabitat.ru/assets/files/publication/GRHS_20.

Union. 2013. *Union of Cities in the Arctic and the High North. Union Analytical Note.* http://krayniy-sever.ru/?page_id=1203.

Union Charter. 1993. *Union of Cities in the Arctic and the High North.* http://krayniy-sever.ru/?page_id=9.

United Nations. 2014. *World Urbanization Prospects, 2014.* http://esa.un.org/unpd/wup/Highlights/WUP2014-Highlights.pdf.

"Upravleniia gosudarstvennoi sluzhby zaniatosti naseleniia Murmanskoi oblasti o dvizhenii rabochei sily na rynke truda Murmanskoi oblasti v ianvare-mae 2015." 2015. *Komitet po trudu i zaniatosti naseleniia Murmanskoi oblasti,* June 2015, www.murman-zan.ru/Attachment.axd?id=2b3df3aa-c84e-4926-9045-1536ff7dd81f.

"VCNG podvodit itogi iubileinogo goda." 2012. *Oblastnaia gazeta,* December 26. Accessed November 24, 2013, www.ogirk.ru/news/2012-12-26/27144.html.

Ventsel, Aimar. 2005. *Reindeer, Rodina, and Reciprocity: Kinship and Property Relations in a Siberian Village.* Vol. 7 of *Halle Studies in the Anthropology of Eurasia.* Berlin: LIT Verlag.

Verdery, Katherine. 1993. "Ethnic Relations, Economies of Shortage, and the Transition in Eastern Europe." In *Socialism: Ideals, Ideologies, and Local Practice,* edited by C. M. Hann, 169–185. London: Routledge.

Verdery, Katherine. 1996. *What Was Socialism, and What Comes Next?* Princeton, NJ: Princeton University Press.

Vlasova, T., and A.N. Petrov. 2010. "Migration and Socio-Economic Well-Being in the Russian North: Interrelations, Regional Differentiation, Recent Trends, and Emerging Issues." In *Migration in the Circumpolar North: Issues and Contexts,* edited by L. Huskey and Ch. Southcott. 163–192. Edmonton: University of the Arctic, Canadian Circumpolar Institute.

Volosova, E.V. 2008. *Vysshee obrazovanie v provintsii: opyt sotsiologicheskikh issledovanii v Ust'-Ilimske (2003–2007).* Ust-Ilimsk.

Ward, Christopher J. 2009. *Brezhnev's Folly: The Building of BAM and Late Soviet Socialism.* Pittsburgh, PA: University of Pittsburgh Press.

Wegren, Stephen K. 1998. *Agriculture and the State in Soviet and Post-Soviet Russia.* Pittsburgh, PA: University of Pittsburgh Press.

Wegren, Stephen K. 2000. "Socioeconomic Transformation in Russia: Where is the Rural Elite'?" *Europe-Asia Studies* 52 (2): 237–271.

Weichhart, P. 2009. "Multilokalität – Konzepte, Theoriebezüge und Forschungsfragen." *Informationen zur Raumentwicklung* 1/2: 1–14.

Wolf, Eric R. 1982. *Europe and the People without History.* Berkeley, CA: University of California Press.

Yakel, Iu.Ia. 2012. "Obshchaia kharakteristika deistvuiuschego zakonodatel'stva. Problemy praktiki primeneniia." In *Sever i severiane. Sovremennoe polozhenie korennykh malochislennykh narodov Severa, Sibiri i Dal'nego Vostoka,* edited by N.I. Novikova and D.A. Funk, 8–21. Moscow: Izdanie IEA RAN.

Yakovleva, Natalia. 2011. "Oil Pipeline Construction in Eastern Siberia: Implications for the Indigenous People." *Geoforum* 42: 708–719.

"Yakutskiie stantsii VSTO obespecheny bespereboinym energosnabzheniiem." 2013. Transneft': Vostoknefteprovod. November 25. Accessed November 27, 2013, http://vostoknefteprovod.ru/presscentr/curnews.aspx?newsID=271.

Yanovskii V.V. 1969. *Chelovek i Sever*. Magadan.

Yarlykapov, Akhmet A. 2008. "Neft' i migratsiia nogaitsev na sever." *Etnograficheskoe obozrenie* 3: 78–81.

Yashunskii, A.D. and N.Y. Zamiatina. 2012. "'Sever kak zona rosta rossiiskoi provintsii." *Otechestvennye zapiski* 50 (5): 227–239.

Yefremov, Igor A. 2014. "Vozrastnye osobennosti migratsionnykh protsessov na Krainem Severe Rossii." http://demoscope.ru/weekly/2014/0581/analit06.php.

Yunusov, Arif. 2003. "Etnicheskie i migratsionnye protsessy v postsovetskom Azerbaizhane. Strategii integratsii v rynok truda i riski." In *Trudovaia migratsiia v SNG. Sotsial'nye i ekonomicheskie efekty*, edited by Zhanna Zaionchkovskaia. Moscow: Tsentr izucheniia problem vynuzhdennykh migratsii v SNG. http://migrocenter.ru/publ/trud_m/06.php.

Zabaykal'skii krai, Department of Federal Migration Service. 2013. *Svedeniya po prebyvaniiu inostrannykh grazhdan iz stran blizhnego zarubezhi'a na territorii Zabaikal'skogo kraia* (s tsel'iu "rabota").

Zamiatina, N.Y. 2012. "Metod izucheniia migratsii molodezhi po dannym sotsial'nykh Internet-Setei: Tomskii gosudarstvennyi universitet kak 'tsentr proizvodstva i raspredelenii' chelovecheskogo kapitala (po dannym sotsial'noi Internet-Seti 'VKontakte.')" *Regional'noe issledovanie* (2): 15–28.

Zamiatina, N.Y. and A.D. Yashunskii. 2012. "Mezhregional'nye tsentry obrazovaniia." *Otechestvennye zapiski* 48 (3): 74–84.

Zhelnina, Z. Iu. 2014. "Turizm Kol'skogo Zapoliar'ia." In *Sovremennye vyzovy i ugrozy razvitiia v Murmanskoi oblasti: regional'nyi atlas*, 42.

Zurabevich, Natalia. *Sotsial'nyi atlas rossiiskikh regionov*. www.socpol.ru/atlas/overviews/social_sphere/kris.shtml.

Index